THE NATURAL HISTORY OF MONITOR LIZARDS

Frontispiece. The author and one of his charges.

Harold F. De Lisle is professor of biology at Moorpark College in Moorpark, California. He has been interested in reptiles since he was very young and has conducted field research on lizards and snakes widely over the western United States and northern Mexico. He served three terms as president of the Southwestern Herpetologists Society, during which time he chaired the Conference on California Herpetology and several large exhibitions of reptiles and amphibians. He is an active member of the Society for the Study of Amphibians and Reptiles.

In recent years he has maintained a large colony of captive varanids which have formed the basis of his own study of monitor lizards. He has studied these lizards in the wild during brief trips to Australia and Indonesia. He is the author of two published books on wildlife, and dozens of articles.

THE NATURAL HISTORY OF MONITOR LIZARDS

Harold F. De Lisle

KRIEGER PUBLISHING COMPANY
MALABAR, FLORIDA
1996

COVER PHOTO
White-throated Monitor (*Varanus a. albigularis*), Transvaal, South Africa.
Photo by K. H. Switak.

Original Edition 1996

Printed and Published by
Krieger Publishing Company
Krieger Drive
Malabar, Florida 32950

Copyright © 1996 by Krieger Publishing Company

All rights reserved. No part of this book may be reproduced in any form or by any means, electronic or mechanical, including information storage and retrieval systems without permission in writing from the publisher.
No liability is assumed with respect to the information contained herein.
Printed in the United States of America.

> FROM A DECLARATION OF PRINCIPLES JOINTLY ADOPTED BY A COMMITTEE OF THE AMERICAN BAR ASSOCIATION AND COMMITTEE OF PUBLISHERS:
>
> This Publication is designed to provide accurate and authoritative information in regard to the subject matter covered. It is sold with the understanding that the publisher is not engaged in rendering legal, accounting, or other professional service. If legal advice or other expert assistance is required, the services of a competent professional person should be sought.

Library of Congress Cataloging-in-Publication Data
De Lisle, Harold F., 1933–
    The natural history of monitor lizards / Harold F. De Lisle.
      p. cm.
    Includes bibliographical references and index.
    ISBN 0-89464-897-7
    1. Monitor lizards. I. Title.
QL666.L29D4 1996
597.95—dc20   95-3063
                                                                                                          CIP
10 9 8 7 6 5 3 2

CONTENTS

LIST OF FIGURES	vii
LIST OF COLOR PLATES	ix
LIST OF TABLES	xi
PREFACE	xiii
CHAPTER 1. INTRODUCTION	1

    Interest in Monitors. Research History.

CHAPTER 2. TAXONOMY AND EVOLUTION	7

    Importance of Taxonomy. Taxonomic History. Phylogeny. Fossil History. Key to Species and Subspecies of *Varanus*.

CHAPTER 3. ANATOMICAL AND PHYSIOLOGICAL SPECIALIZATIONS OF VARANIDS	25

    Vomeronasal System and Olfaction. Vision. The Ear and Hearing. Epidermal Senses. Tongue and Hyoid Apparatus. Teeth and Jaws. Digestive Tract. Locomotion. Respiration. Circulation. Water and Salt Balance. Monitor Energetics. Reproductive Specializations.

CHAPTER 4. ECOLOGY AND BEHAVIOR	53

    Diet. Foraging Behavior. Courtship and Mating. Male Combat Behavior. Mating Systems. Nesting. Eggs. Hatching and Hatchlings. Growth. Longevity. Population Dynamics. Activity Patterns. Thermal Behavior and Ecology. Habitat Utilization. Defensive Behavior. Predation. Intelligence.

CHAPTER 5. CAPTIVE MANAGEMENT	89

    Obtaining Healthy Specimens. Housing. Feeding. Water. Breeding. Disorders and Diseases.

CHAPTER 6. CONSERVATION	107

    Endangered and Threatened Species. Threats to Monitors. Regional Problems. Conservation Laws and Education.

COLOR PLATES FOLLOW PAGE 114	
CHAPTER 7. SPECIES AND SUBSPECIES OF THE GENUS *VARANUS*	115
CHAPTER 8. THE FUTURE FOR VARANIDS	173

    Knowledge Gaps. Contributions from Captive Studies.

APPENDIX A. RECOGNIZED TAXA OF THE GENUS *VARANUS*	177

    Cross-Referenced by Scientific Name.

APPENDIX B. RECOGNIZED TAXA OF THE GENUS *VARANUS*	179

    Cross-Referenced by Common Name.

REFERENCES	181
SCIENTIFIC NAME INDEX	193
GENERAL INDEX	197

LIST OF FIGURES

Frontispiece. The author and one of his charges.

Figure 2.1 Skull of *Varanus salvator*

Figure 2.2 Diagram of the three main radiations within the genus *Varanus*

Figure 2.3 Proposed phylogeny of the genus *Varanus*, based on morphology of the hemipenis

Figure 2.4 Proposed phylogeny of the genus *Varanus*, based on lung morphology

Figure 2.5 Proposed phylogeny of the genus *Varanus*, based on chromosome morphology

Figure 2.6 Proposed phylogeny of the genus *Varanus*, based on microcomplement fixation

Figure 2.7 Phylogeny of the African radiation of *Varanus*

Figure 2.8 Phylogeny of the Indo-Asian radiation of *Varanus*

Figure 2.9 Phylogeny of the Australian radiation of *Varanus*

Figure 2.10 Proposed radiation of the genus *Varanus* during the Miocene epoch

Figure 3.1 Generalized anatomy of the nose region of *Varanus*

Figure 3.2 Field of vision for a typical varanid

Figure 3.3 Anatomy of the major features of the eye of *Varanus*

Figure 3.4 Scleral ossicles of the eye of *Varanus*

Figure 3.5 External ear openings of various species of varanids

Figure 3.6 Cross (coronal) section through the ear region of *Varanus*

Figure 3.7 Typical dorsal scale pattern of *V. albigularis*

Figure 3.8 Mouth and tongue region of *Varanus*

Figure 3.9 Bones of the hyoid apparatus of *Varanus*

Figure 3.10 Maxillary tooth of *Varanus komodoensis*

Figure 3.11 Internal anatomy of a varanid lizard

Figure 3.12 Respiratory and urinary anatomy of a varanid lizard

Figure 3.13 Internal structure of the varanid lung

Figure 3.14 Diagram of the internal structure of the varanid heart

Figure 3.15 Diagram of the varanid kidneys

Figure 3.16 Sagittal section through the varanid cloacal region

THE NATURAL HISTORY OF MONITOR LIZARDS

Figure 3.17 Reproductive structures of a male varanid

Figure 3.18 Hemipenis of some representative varanids

Figure 3.19 Reproductive structures of a female varanid

Figure 4.1 Copulation in *Varanus komodoensis*

Figure 4.2 Copulation in *Varanus rosenbergi*

Figure 4.3 Copulation in *Varanus olivaceus*

Figure 4.4 Copulation in *Varanus timorensis*

Figure 4.5 Ritual combat posture between male *Varanus varius*, showing neck-arching behavior

Figure 4.6 Ritual combat posture between male *Varanus gilleni*, showing clasping embrace

Figure 4.7 Termite mound nest of *Varanus rosenbergi*

Figure 4.8 Bank soil nest of *Varanus bengalensis*

Figure 4.9 Eggs of *Varanus varius*

Figure 4.10 Internal structure of a developing varanid egg

Figure 4.11 Recently hatched varanid eggs, showing slits made by egg-tooth of the baby lizard

Figure 4.12 Karyotypes of male (top) and female (bottom) *Varanus varius*

Figure 4.13 Bipedal defensive posture in *Varanus gouldii*

Figure 4.14 (A) Normal walk posture, (B) threat walk posture

Figure 5.1 Cage setup for a terrestrial monitor

Figure 5.2 Cage setup for an arboreal monitor

Figure 5.3 *Varanus a. albigularis* with tick infestation

LIST OF COLOR PLATES

Cover		*V. a. albigularis*
Plate 4.1.		*V. varius* mating
Plate 4.2	a.	*V. albigularis*
	b.	*V. d. dumerilii*
	c.	*V. niloticus*
	d.	*V. spenceri*
	e.	*V. t. timorensis*
Plate 7.1	a.	*V. a. brachyurus*
	b.	Habitat, *V. a. brachyurus* and *V. giganteus*
Plate 7.2	a.	*V. a. albigularis*
	b.	Habitat, *V. a. albigularis*
Plate 7.3.		*V. a. ionidesi*
Plate 7.4.		*V. a. microstictus*
Plate 7.5.		*V. beccarii*
Plate 7.6.		*V. b. bengalensis*
Plate 7.7.		*V. b. irrawadicus* (holotype)
Plate 7.8.		*V. b. nebulosus*
Plate 7.9.		*V. b. vietnamensis* (holotype)
Plate 7.10.		*V. brevicauda*
Plate 7.11.		*V. caudolineatus*
Plate 7.12.		*V. d. doreanus*
Plate 7.13.		*V. d. dumerilii*
Plate 7.14.		*V. eremius*
Plate 7.15.		*V. exanthematicus*
Plate 7.16.		*V. flavescens*
Plate 7.17	a.	*V. f. flavirufus*
	b.	Habitat, *V. flavirufus* and *V. gilleni*
Plate 7.18.		*V. f. gouldii*
Plate 7.19	a.	*V. giganteus*
	b.	*V. giganteus*
Plate 7.20.		*V. gilleni*
Plate 7.21.		*V. glauerti*
Plate 7.22.		*V. glebopalma*
Plate 7.23.		*V. g. gouldii*
Plate 7.24.		*V. g. horni*
Plate 7.25.		*V. g. griseus*
Plate 7.26.		*V. g. caspius*
Plate 7.27.		*V. g. koniecznyi*
Plate 7.28	a.	*V. indicus*
	b.	Habitat, *V. indicus* and *V. prasinus*
Plate 7.29.		*V. jobiensis*
Plate 7.30	a.	*V. kingorum*
	b.	Habitat, *V. kingorum* and *V. glebopalma*
Plate 7.31.		*V. komodoensis*
Plate 7.32.		*V. mertensi*
Plate 7.33	a.	*V. mitchelli*
	b.	Habitat, *V. mitchelli*
Plate 7.34	a.	*V. niloticus*
	b.	Habitat, *V. niloticus*
Plate 7.35.		*V. olivaceus*
Plate 7.36.		*V. pilbarensis*
Plate 7.37.		*V. prasinus*
Plate 7.38.		*V. primordius*
Plate 7.39	a.	*V. rosenbergi*
	b.	Habitat, *V. rosenbergi*
Plate 7.40.		*V. rudicollis*
Plate 7.41.		*V. salvadorii*
Plate 7.42.		*V. s. salvator*
Plate 7.43.		*V. s. cumingi*
Plate 7.44.		*V. s. marmoratus*
Plate 7.45.		*V. s. nuchalis*

THE NATURAL HISTORY OF MONITOR LIZARDS

Plate 7.46   a. *V. scalaris*
               b. Habitat, *V. spenceri*,
                   V. storri, and *V. tristis*

Plate 7.47   a. *V. spenceri*
               b. Habitat, *V. spenceri*

Plate 7.48. *V. spinulosus*

Plate 7.49. *V. s. storri*

Plate 7.50. *V. teriae*

Plate 7.51. *V. t. timorensis*

Plate 7.52. *V. t. similis*

Plate 7.53. *V. t. orientalis*

Plate 7.54   a. *V. varius*
               b. *V. varius* (banded phase)
               c. Habitat, *V. varius*

Plate 7.55. *V. yemenensis*

LIST OF TABLES

Table 2.1. Derived Character States for the Family Varanidae

Table 3.1. Volumetric Relations of Parts of the Digestive Tract of Varanid Lizards

Table 3.2. Degree of Arboreality in *Varanus* Species

Table 3.3. Comparative Metabolic Rates of Some Varanid Lizards

Table 4.1. Importance Indexes of Various Prey Types (by volume percent) of Monitor Species

Table 4.2. Summary of Courtship and Mating Behavior in *Varanus*

Table 4.3. Summary of Data on Eggs of *Varanus*

Table 4.4. Length and Weight of Hatchling Varanids

Table 4.5. Average Linear Growth Rate of Hatchling Varanids during First Year

Table 4.6. Longevity Records for Captive Varanids

Table 4.7. Mean Home Ranges for Adults of Some Varanid Species

Table 4.8. Daily Activity Behavior Patterns of Some Varanids

Table 4.9. Activity Temperatures of Varanid Lizards

Table 5.1. Suggested Minimum Nutrient Levels in Diet of Captive Varanids.

Table 5.2. References of Captive Reproduction in the Genus *Varanus*

Table 6.1. Estimated Annual Take in Monitor Skins for International Trade

Table 6.2. Minimum Net Trade in Live Varanids Recorded in CITES Annual Reports 1981-1990.

PREFACE

Why a book about monitor lizards? The exotic nature of monitor lizards has added to the fascination for them in the Western world, although our review of the stories and beliefs about monitors in their native lands indicates this fascination by humankind is nearly universal. Many unique aspects of their biology undoubtedly contribute to this fascination. This unique biology is the topic of this book.

I became acquainted with monitors as a college student when some of the first juvenile Bengal monitors were imported from Thailand. After paying the exorbitant price of five dollars for one of these unusual lizards, I established it in a terrarium. In those days before full-spectrum lighting, heating stones or care manuals, I was lucky that it survived as long as it did. I have been fascinated with these lizards ever since.

This book is the result of many years' experience in keeping various varanids in captivity, observing their behavior, and keeping careful notes. I have had two opportunities to study monitors in the wild, albeit too briefly—in Australia and Indonesia.

I am indebted to other researchers for most of the information in this book. I have attempted to compare the vast amount of data appearing on many aspects of varanid biology and draw some generalizations from it and from my own experience. Most of the published information is contained in scientific journals and monographs, and appears in several languages. This book is meant to be a review of this information and to help the nonscientist understand the place of monitors in nature.

I am grateful to the colleagues who reviewed and criticized various draft chapters of the book. I acknowledge the many helpful comments of G. R. Stewart, S. R. Goldberg, R. W. Van Devender, and W. Auffenberg. However, the responsibility for any errors of fact or grammar are mine.

I am also indebted to several people who helped me directly in my study of varanids—F. Yuwono, C. and G. Roscher, H. Fisher, R. L. Bezy, D. R. King, and D. Bennett.

Credit for photography goes to D. Northcutt, W. Bohme, M. Gaulke, K. Tepedelen, R. W. Van Devender, G. Visser, M. Bayless, R. Hoser, K. Switak and Yang Datong.

Chapter 1
INTRODUCTION

Varanid lizards have fascinated herpetologists for a long time. Monitors are different from most other lizards. They have anatomical and physiological characteristics that make them more like birds and mammals than less active lizards. Their ecology is just beginning to become known, thanks especially to Walter Auffenberg (1981a, 1988, 1994). Behavioral studies indicate that they are the most intelligent of all lizards and probably of all reptiles (De Lisle, unpubl. data). Many monitor lizards are large and impressive, and often form major exhibits in zoos.

The family Varandiae includes the largest living lizards. It is an ancient saurian lineage which diverged from the mososaurs (marine lizards) in the Early Cretaceous (100 million years ago). Their resemblance in some anatomical respects to snakes is now generally believed to be a case of convergent evolution rather than indicative of any close ancestry. Although monitors may have originated in North America, they spread into the Old World and today are found through all of Africa, east through southern Asia to Australia and the Solomons (Map 1.1). There are some 70 species and subspecies described to date (and several more forms awaiting formal description). Monitors, however, lack the variety of body forms found in most lizard families. Ranging in size from 0.32m to nearly 3.5m in total length, size is the most variable anatomical characteristic in the family. They vary in mass by five orders of magnitude. No other terrestrial animal genus shows such size variation; there is proportionately almost as much difference in mass among species of varanids as there is between a mouse and an elephant. Many monitors have limbs that are thick and muscular, and each foot has five toes equipped with strong, curved, sharp claws. Yet with minimal morphological variation they have diversified ecologically, being found from deserts to tropical rain forests. All, even the desert forms, are excellent climbers. Their size variability has led to a diet so varied as to be as small as an ant or large as a buffalo. In some areas, such as northern Australia, as many as 10 species occur sympatrically (live in the same place). Where this occurs the species usually range in size from small to large, with the smaller forms forming part of the food base for the larger forms (Pianka, pers. comm.)

INTEREST IN MONITORS

The appeal of monitors to amateurs and professional herpetologists is undeniable. *V. komodoensis*, as the world's largest lizard, has an appeal to the general public so great that it is the central tourist attraction of the Lesser Sunda Islands of Indonesia. In parts of their range, monitors have acquired a fascinating folklore. In Hindu lore the monitor is known as "biscobra," implying that it is twice as deadly as the cobra, and people flee at the sight of it. It is

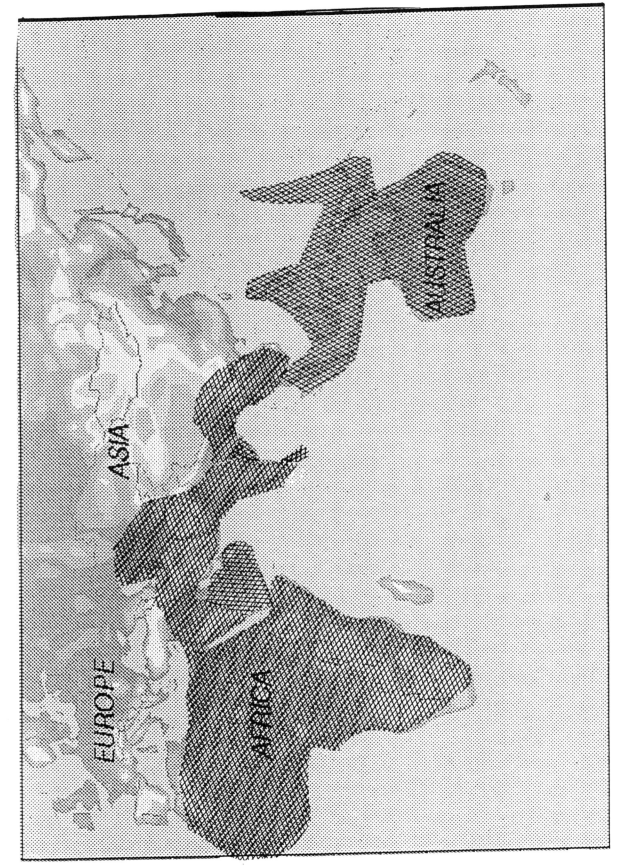

Map 1.1. Range of the genus *Varanus*.

reputed to fly through the air attacking intruders into its domain, and its bite is instantly fatal. Monitor bites seem to be considered fatal among many Asian cultures, perhaps because they so often carry serious infection. Goannas, as monitors are called in Australia, figure large in the aboriginal folklore or "dreamtime." Elsewhere, fresh monitor meat used as a poultice is considered essential to cure an open wound.

Blood from monitors is given to growing boys in Sri Lanka in the belief that it gives them strength. In parts of Asia (Thailand, Borneo, Malaysia) monitors take on the symbolism that black cats do in western culture—symbols of bad luck if one crosses the path in front of you. There is also a belief in Malaysia that witches can change into monitors at will.

Monitor skin, tanned and dyed and finished to a high gloss, is tough, thin and decorative. It is much in demand in the fashion business for fancy leather articles like watch straps, purses, and shoes.

Among amateur reptile lovers, most monitor specimens command a high price ($100–$1,000) for legal specimens. There is even a Varanid Information Exchange with a bimonthly newsletter.

Among professionals, the late Robert Mertens ranks as the premier researcher of varanids. He confessed in 1942 to first becoming captivated by his future research topic by a Nile monitor he had in a terrarium as a boy. He wondered at its "dignity, physique, power and intelligence." Walter Auffenberg is the greatest living scholar of varanids. His devotion has led him to spend months on end camped out in often inhospitable Asian environments to study his favorite animals.

There is considerable interest in the captive maintenance of monitors. There exist two booklets (Balsai, 1992, Anon., 1992) on the care and maintenance of savannah monitors, the most commonly kept species. Monitors are also popular in zoos, with the water monitor and Nile monitor, which reach impressively large size, being most commonly exhibited. Monitors are popular because of their relatively large size among lizards. Some have attractive color patterns, another key to popularity. Their catholic diets make them fairly easy to maintain in captivity, at least as far as providing food is concerned. After one gets to know monitors, their intelligence and so-called dinosaur-like habits reinforce their popularity. A couple of species, e.g., *V. exanthematicus* and *V. dumerili*, can even be tamed so that they can be handled; however, most species do not make good pets. If approached or handled they lash out with the tail, hiss, struggle and claw at the handler, defecate, and bite. They are best admired from behind a pane of glass.

Captive breeding has not been very successful either privately or in zoos. As of 1989 (Horn and Visser), just 18 species had reportedly been captive bred; and only one (*V. acanthurus*) had been raised and bred through more than one generation. Literature reports of captive breeding for the past five years have been few. The smaller species have yielded the best results so far, indicating that considerable space may be needed for the essential courtship rituals to occur among what are usually solitary animals. Specimens originating from the same geographical area which have been reared in captivity seem to make the most reliable breeding pairs. Separation of the sexes appears to be necessary to induce breeding.

RESEARCH HISTORY

Although monitors have been known to science since the time of Albert Seba (1734), research was limited to taxonomy until the second half of the nineteenth century.

Albert Gunther, German M.D. and curator at the British Museum, wrote his *Anatomy of Varanus niloticus* in 1861, ushering in the age of more detailed monitor research. Wilhelm Peters, German zoologist and African explorer, published the first natural history account of African monitors in 1870. W. Rathgen authored the first account on captive management in 1894, on *V. griseus*.

The first quarter of the twentieth century saw an increasing number of studies of the most common species. In 1903 Einar Loonberg described the diet of *V. niloticus*, and J. Scherer (1907) studied the natural history of the same species in Senegal. Lorenz Muller wrote the first captive care manual for the savannah monitor in 1905.

Raymond Ditmars, in his 1910 classic, *Reptiles of the World*, devotes several pages to two species with which he had experience in captivity (*V. griseus* and *V. salvator*). In 1912 P. A. Ouwens reported the existence of a large monitor in the Lesser Sunda Islands of the Dutch East Indies. W. Douglas Burden led an expedition in 1926 to Komodo and verified the existence of *V. komodoensis*. Many studies of different aspects of the biology of this species were made in the decade following.

Raymond Cowles authored a life history of *V. niloticus* in 1928–30. H. C. Smith published his study on the monitor lizards of Burma in 1931.

Modern research on monitor lizards began with the famous German herpetologist Robert Mertens. Born in 1894 in St. Petersburg, Russia, of German parents, Mertens is considered the dean of modern European herpetologists (Adler, 1989). His boyhood interest in reptiles was encouraged by his father, and as mentioned previously he kept a captive Nile monitor for several years. In 1919 Mertens joined the staff of the Senckenberg Museum in Frankfurt where he remained until his retirement in 1960. Besides his interest in systematics, Mertens was always interested in studying living animals, especially reptile behavior. He kept a large collection of live specimens at his home. He contributed regularly to German terrarium journals and his encouragement is held largely responsible for the large and active group of amateur herpetologists in Germany.

It was in the depths of World War II (1942) that Mertens published his three volume monograph, *Die Familie de Warane (Varanidae)*. This is still the basic authority on varanid systematics.

The last 50 years have seen an increasing research interest in monitors. Beyond doubt, the outstanding researcher of varanids in the last half of the twentieth century is American Walter Auffenberg (1928–). Based at the University of Florida, Gainesville, Auffenberg came to monitors rather late in his career. He spent 13 months in the Lesser Sunda Islands (1969–71) which resulted in his book *The Behavioral Ecology of the Komodo Monitor*, considered a pioneer study in the new field of behavioral ecology. This was followed by a 22-month (1976–83) field study on Luzon, Philippines, resulting in an another behavioral ecology monograph (1988) on Gray's Monitor Lizard, revealing the only regularly frugivorous (fruit-eating) lizard in the world. Auffenberg has culminated his studies with a third behavioral ecology study (1994), this time on the Bengal monitor, which he studied at various time and places from 1974 to 1987.

Among the other recent research studies of monitors should be mentioned the physiological research of Brian Green and his associates on several Australian species. Eric Pianka has also contributed to the natural history of several of Australia's desert species. These scientists and many others have followed Glenn Storr in making the biology of Australian monitors now among the best known.

Maren Gaulke did an ecological field study (1989) of the water monitor in the Philippines. J. Losos and H. Greene completed an important survey of the diet of wild monitors (1988).

Most research on monitors continues the efforts at clarifying their systematics. Some of these researchers should be mentioned: Dennis King (Australia), Wolfgang Böhme (Germany), and George Sprackland (U.S.). The Second World Congress of Herpetology in 1993–94 sponsored a symposium on the biology of monitor lizards.

Chapter 2
TAXONOMY AND EVOLUTION

IMPORTANCE OF TAXONOMY

Taxonomy has as a major objective to sort out closely related organisms and assign them to separate species, describing the diagnostic characteristics that distinguish species from one another. Of all taxonomic units (family, genus, species, etc.), only the species exists as a biologically cohesive unit. All other branches of organismal biology depend on correct taxonomy to support their conclusions. The first attempts at diagnosing species of varanids were done almost entirely from preserved specimens, often very few. As studies increased in number and depth over the last century, species of varanids have frequently been rediagnosed with the result of an over all increase in the number of species. A synopsis of these historical changes is included under the species' descriptions in chapter 7. Taxonomy has as another important objective to describe the discovery of new species. Ten new species of varanids have been discovered in the last 50 years. The other new species that have been added have been the result of rediagnosis of previously described specimens.

There are two basic rules that must be followed when describing or redescribing a species. The first is the law of priority, which states that the valid name of a species can be only that name under which it was first designated, providing certain conditions are met. Two of these conditions are (1) the name has not been used before for some other animal; (2) the name was published with a description of the characters used in diagnosis. The second rule is the law of synonymy. Synonyms are different names for the same species. The oldest available name is the valid name, often called the senior synonym, in contrast to junior synonyms, which are more recent and therefore invalid names. Much of the taxonomic history of varanids for the last fifty years has involved applying these two rules to descriptions originally made in the nineteenth and early twentieth centuries.

For example, in 1951 Robert Mertens described a new monitor from New Guinea and named it *Varanus karlschmidti* after his friend Karl Schmidt, the dean of American herpetologists for many years (Adler, 1989). In 1991, Wolfgang Bohme discovered that this monitor was identical to one described by Ernst Ahl in 1932 as *Varanus indicus jobiensis*. Thus the older name takes precedence. When Böhme recognized that this form was a species separate from the mangrove monitor (*V. indicus*), he had to use the senior synonym (*jobiensis*) to redescribe it.

TAXONOMIC HISTORY

Monitors have had a long and tortuous treatment in the systematic literature. The first taxonomic designation of monitor lizards dates to the tenth edition of Linnaeus's *Systema Naturae* (1758), which lists *Lacerta monitor* (now *V. niloticus*). In 1802-3 Francois Daudin, a French amateur naturalist, published his influential *Histoire Naturelle, Generale et Particuliere des Reptiles*. He included seven modern monitor species (*albigularis, griseus, salvator, niloticus, bengalensis, varius,* and *exanthematicus*) in the genus *Tupinambis*.

In 1820 Blasius Merrem, a professor at the University of Marburg, Germany, introduced the genus name *Varanus* in his *Tentamen Systematis Amphibiorum*. *Varanus* is a latinization of the Arabic "waran," the Egyptian name for the Nile monitor. The word literally means "monitor" from the ancient belief that these lizards would alert people to the presence of crocodiles.

John E. Gray, keeper of the zoological collections at the British Museum, placed these lizards in the family Varanidae (1827) with the following diagnosis: "Tongue retractable; head and body scaly, femoral pores none; palate toothless." The definition of the family Varanidae recognized today is that provided in 1885 by George Boulenger, who was also curator at the British Museum: "Head covered with small juxtaposed scales, body with even smaller scales, and each scale surrounded by a rosette of tiny granular scales (Fig. 3.7); tongue very long, slender, deeply bifid, retractile into a sheath; ventrals quadrangular, arranged transversely; small pits on scales; osteoderms present; tail elongate, not fragile; pupil round; eyelid well developed."

In 1942 Robert Mertens published his three volume monograph, *Die Familie der Warane (Varanidae)*, which included the first complete classification of living monitors. Mertens's work was severely hampered by World War II which precluded access to specimens in many countries. Nevertheless, his work has remained the foundation of monitor taxonomy for 50 years. Of course, a number of modifications have been made by subsequent studies. Mertens original classification placed 24 species in eight subgenera. He also recognized subspecies in most of the major species. Mertens's classification was based on anatomy (mainly osteology) and external features such as scalation and coloration. The status of his subgenera has been debated intensely ever since with little agreement, and monophyly of the subgenera is in serious doubt (Baverstock, 1993). I omit the use of this taxonomic category in this work.

This work makes use of the subspecies concept of geographically distinct populations. There is increasing evidence that in many (most?) cases these distinctions have validity in the natural world. Several varanids have been shown to prefer mating with a member of their own subspecies over an individual of another subspecies, and are more likely to produce viable eggs (cf. Auffenberg, 1994).

The overall place of monitors in the scheme of vertebrate taxonomy is shown below (modified from Gauthier et al., 1988; Estes et al., 1988).

Amniota
  Mammalia
  Reptilia
     Chelonia (turtles)
     Sauria
        Archosauria (ruling reptiles)
           Dinosauria (dinosaurs)
           Aves (birds)
           Crocodylia (crocodilians)
        Lepidosauria
           Rhynchocephalia (*Sphenodon*)
           Squamata (snakes, lizards, and amphisbaenians)
              Iguania (iguanas, agamas, chameleons)
              Scleroglossa (all other lizards)
                 Anguimorpha (xenosuarids, anguids,
                    heloder-matids, varanids)
                    Varanoidea
                       Helodermatidae
                       Varanidae
                          Lanthanotus
                          Varanus

The scales of varanids are mostly small, pebble-like granules which may form rings around larger, juxtaposed scales with conspicuous pits leading to tactile organs (Fig. 3.7). The head scales are polygonal; body and appendage scales oblong or squarish. There are thin disks of bone called osteoderms in some scales. The distinctive body form includes a long neck in many species. The head is usually very long and frequently has a pointed snout. A skin fold is present across the throat. The limbs are strongly built, five-toed and have long, strong, curved claws. The tail is long and muscular and does not break easily. The large eyes have almost circular pupils. The ears are external with an exposed opening. Femoral and preanal glands are absent. Table 2.1 lists the character states that have evolved in and identify the family Varanidae.

V. salvator has the fewest derived (i.e., the most primitive) characters and is therefore believed to resemble most closely the ancestral *Varanus*.

PHYLOGENY

Evolutionary biologists are keenly interested in the evolutionary pathways that have been followed during the history of a group of animals. Such a

Table 2.1
Derived Character States for the family Varanidae (modified from Greer, 1989)

I. External morphology
  Head scales fragmented

II. Skull, mandible, and hyoid morphology (Fig. 2.1)
  Premaxillae fused, at least dorsally
  Nasal process of premaxillary long, extending to frontal
  Narial slit extends posteriorly to separate completely nasal from maxillary and prefrontal
  Frontals encircle the forebrain
  Parietals fused
  Lacrimal foramen divided
  Prefrontal and jugal widely separated by lacrimal
  Alar process of prootic overlaps descending process of parietal
  Foramen between premaxilla and maxilla in palate
  Maxilla contacts vomer behind opening to vomeronasal organ
  Vomers lack lateral shelves
  Ectopterygoid contacts palatine to exclude maxilla from infraorbital vacuity
  Palatines lack choanal grooves
  Palatal foramen entirely within palatine
  Teeth pointed, recurved, basally fluted
  Lower jaw hinged medially
  Second branchial arch absent

III. Postcranial skeletal morphology
  Presacral vertebrae = 27
  Lumbar vertebrae = 1
  Vertebrae with zygosphenes
  Caudal vertebrae lack autotomy septa; tail does not regenerate
  Centra of caudal vertebrae with pedicels for caudal haemapophyses
  First rib on fifth presacral vertebra
  Inscriptional ribs incomplete
  Clavicles rod-shaped, medially separated
  Interclavicle anchor-shaped
  Mesosternum absent
  Intermedium absent
  Fifth metatarsal separated from fourth metatarsal
  Hemipenis contains two bones

IV. Soft anatomy
  Tongue long and thin, deeply forked apically
  Intrapulmonary bronchi cartilage-lined
  Lungs with extensive faveoli

V. Behavior
  Males perform ritual combat

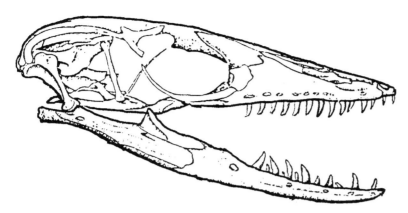

Figure 2.1. Skull of *Varanus salvator*.

pathway is called a phylogeny, and its illustration a phylogenetic tree. Phylogenies are fundamental to comparative biology; there is no doing it without taking them into account. However, establishment of phylogenetic relationships is very difficult. Evidence is usually inconclusive on many points and opinions differ even (and especially) among the experts best acquainted with the group.

Taxonomic data is used for constructing phylogenies. A phylogenetic study is based on a tabulation of shared characters. Not only anatomical features should be tabulated, but also biochemical, genetic, ecological, physiological, and geographic data to the extent to which they are available. In spite of the subjective nature of this kind of study, phylogenetic diagrams are useful summarizations of taxonomic knowledge and provide a pictorial representation of the author's concept of the evolutionary history of the group. Such a simple diagram may show more than many pages of detailed description.

Phylogenetic trees have two significant aspects: (1) the location of branch points along the tree, symbolizing the relative time of origin of different species, and (2) the degree of divergence between branches, representing how different two species have become since branching from a common ancestor. If phylogeny is to be based on evolutionary history, which property of phylogenetic trees should be given the greatest weight when grouping species? This question continues to divide systematics into two schools of thought—phenetics and cladistics.

Phenetics makes no phylogenetic assumptions and decides affinities on the basis of measurable similarities and differences. As many anatomical characteristics as possible are compared with no attempt to sort homology from analogy. Critics of this school argue that overall phenotypic similarity is not a reliable index of phylogenetic relationship. Cladistics classifies species according to the order in time that branches arise along a phylogenetic tree, excluding the degree of divergence from consideration. Critics of this school argue that the extent of morphological divergence should be included.

A third group of systematists, calling themselves classical evolutionary biologists, have attempted to balance the criteria of phenetics and cladistics and where a conflict arises a subjective judgment is made about which type of information should be given higher priority.

Therefore, it is not uncommon to find more than one phylogenetic scheme for a given group of animals, each being an hypothesis or model. In recent years several phylogenies for varanid lizards have been proposed based on different sets of characteristics (anatomical, biochemical, or genetic) investigated. It is hoped that additional evidence will eventually allow us to reject those phylogenies that are incorrect.

Most authorities agree on three main radiations in the evolution of the varanids. These are centered in Africa, Indo-Asia, and in Australia (Fig. 2.2). Each of these primary radiations appears to have a major secondary radiation.

The following cladograms represent phylogenies proposed by various authorities. The closer the names on the right, the closer the relationship. The distance from branching on the left indicates the relative date of separation of the line of evolution from common ancestry. Remember critics of cladistics would claim that morphological variation may not be correlated with time. They say we have no clock, and the real clocks may be different on different branches of the tree. See Figs. 2.3–2.6.

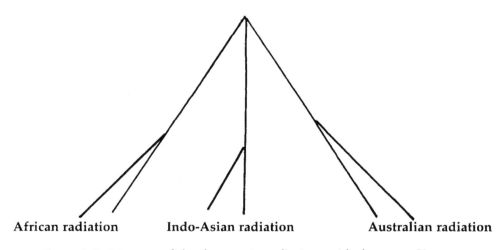

Figure 2.2. Diagram of the three main radiations with the genus *Varanus*.

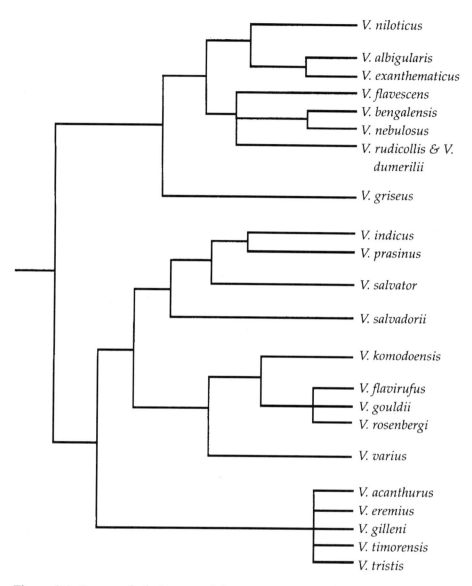

Figure 2.3. Proposed phylogeny of the genus *Varanus*, based on morphology of the hemipenis, after Böhme (1988).

THE NATURAL HISTORY OF MONITOR LIZARDS

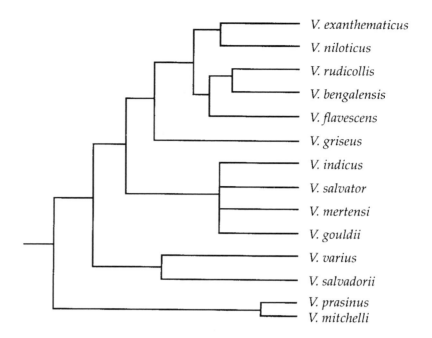

Figure 2.4. Proposed phylogeny of the genus *Varanus*, based on lung morphology, after Becker et al. (1989).

Comparison of different characters almost invariably produces different cladograms because different sorts of characters evolve at different rates, with the results illustrated above. The cladograms in Figs. 2.7–2.9 are my attempt at a synthesis of phylogenies proposed by the above authorities and others.

FOSSIL HISTORY

Varanid-like lizards (aigialosaurids, dolichosaurids, varanoids) first appear in the fossil record during the Lower Cretaceous, about 100 million years ago. Fossils from Mongolia, from the Upper Cretaceous (*Telmasurus, Saniwides*, and *Cherminotus*), are the earliest known terrestrial varanoids (Carroll and Debraga, 1992). The climate in Mongolia at the time is believed to have been as arid as today but somewhat warmer. *Telmasurus* is also known from North America. By the time of the break-up of Pangea (75 million years ago), the ancestors of *Varanus* had spread perhaps to much of that supercontinent. However, the absence of Cretaceous fossils from South America and Africa lead most authorities to agree on a Laurasian origin of the varanids (Pregill et al., 1986). The earliest fossil that can be definitely ascribed to the family Varanidae is from the Upper Cretaceous deposits in Mongolia (*Saniwides mongoliensis*), which may have fed on dinosaur eggs (Pianka, 1994). Fossils of large varanids appear associated with Upper Cretaceous dinosaur nests in

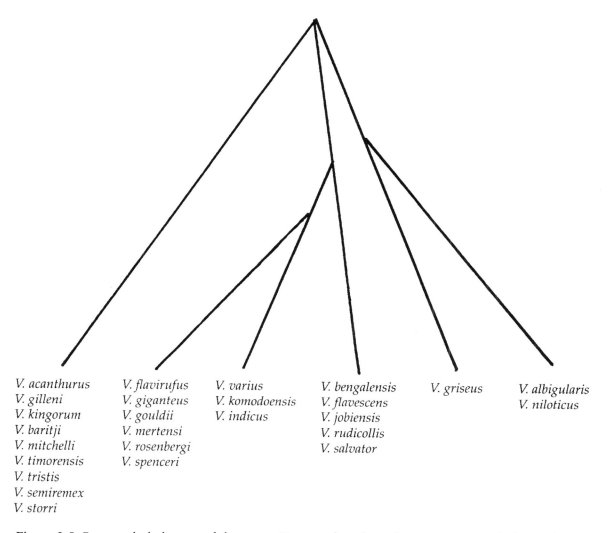

Figure 2.5. Proposed phylogeny of the genus *Varanus*, based on chromosome morphology, after King (1990).

Montana (Horner, 1987). The fondness for eggs for which many varanids are noted may have become established early in their evolution.

Saniwa, from the Eocene of North America, is the earliest member of the varanids for which an entire skeleton is known. It is similar to the modern genus *Varanus* in most respects.

The genus *Varanus* first appears, more or less simultaneously, in fossil deposits of Miocene age (15 million years ago) in eastern Europe (*V. hofmanni*), Africa, and Australia. It seems likely that the genus arose in Asia and then radiated into Africa and Australia (King and King, 1975; King and

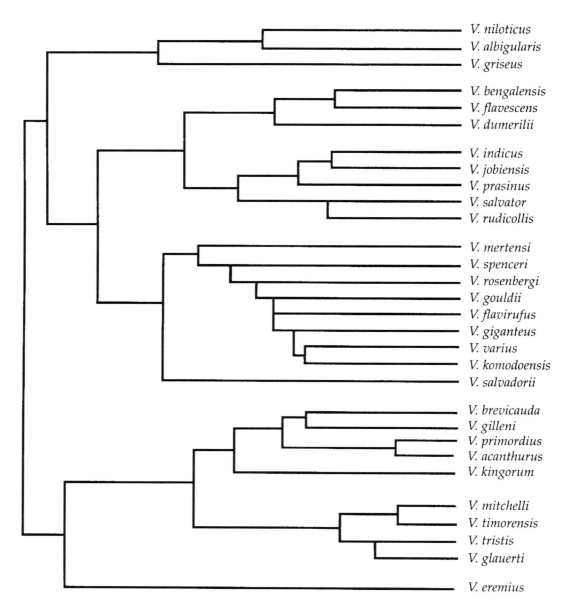

Figure 2.6. Proposed phylogeny of the genus *Varanus*, based on microcomplement fixation (MCF) of albumin, after Baverstock et al. (1993).

Green, 1993) (Fig. 2.10). The Miocene was the last epoch when tropical and subtropical climates dominated most continental regions.

The most famous fossil monitor is usually placed in a separate genus, but it is certainly closely related to modern varanids. *Megalania prisca* is the largest terrestrial lizard known, reaching a total length of 8 meters and an estimated weight of 600 kg. It is known from the Pleistocene of Australia. The

Taxonomy and Evolution

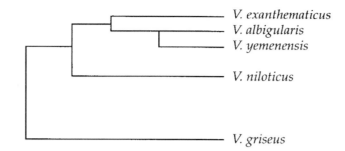

Figure 2.7. Phylogeny of the African radiation of *Varanus*.

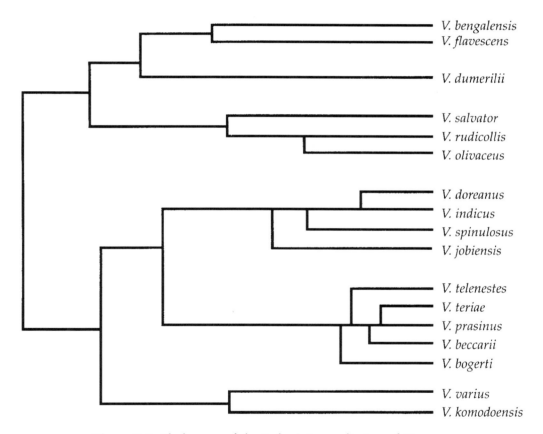

Figure 2.8. Phylogeny of the Indo-Asian radiation of *Varanus*.

largest terrestrial carnivores native to Australia today are monitors. It is not surprising that a giant carnivorous monitor evolved during the Pleistocene, when Australian fauna also included numerous large herbivorous marsupials. Fossil beds on the island of Timor have also yielded bones of an extinct large monitor (Hooijer, 1972). A race of pygmy elephants once lived on some of

THE NATURAL HISTORY OF MONITOR LIZARDS

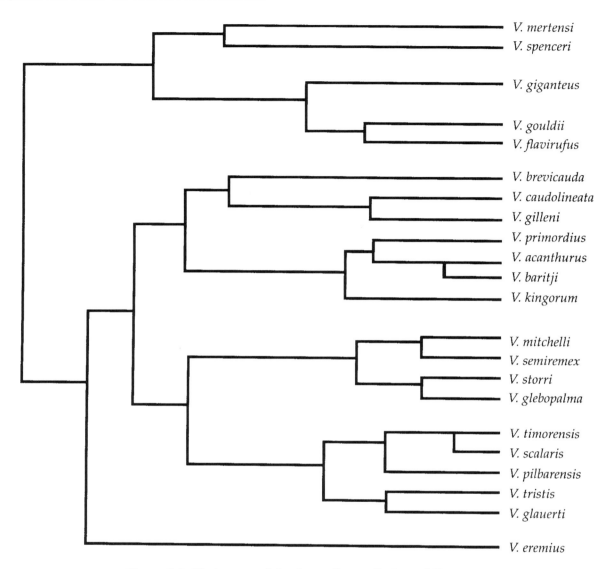

Figure 2.9. Phylogeny of the Australian radiation of *Varanus*.

these Indonesian islands and may have formed the original food base for both extinct and living giant monitors.

In Recent time monitors have continued to flourish so that today there are 46 species presently described. The greatest center of diversity remains Australia.

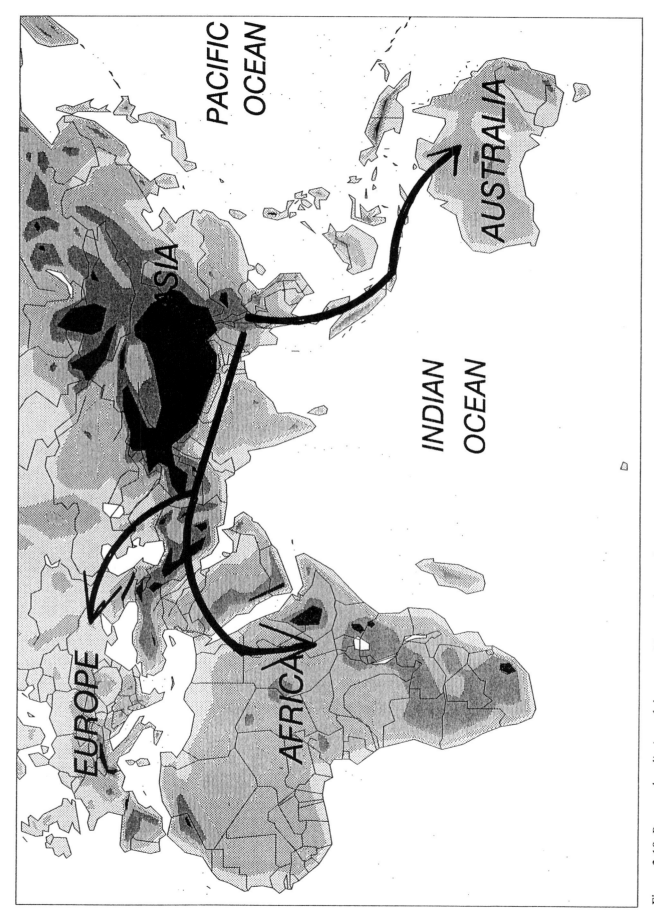

Figure 2.10. Proposed radiation of the genus *Varanus* during the Miocene epoch.

KEY TO SPECIES AND SUBSPECIES OF *VARANUS*

This key emphasizes external features of living lizards and should be useful to a broad spectrum of herpetologists, including amateurs and museum curators. Color patterns refer to adult lizards; juveniles as a rule are differently and more vividly patterned and colored.

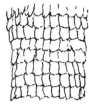

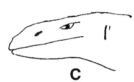

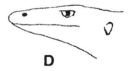

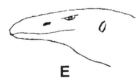

1A. Tail strongly compressed laterally, with distinct median double keel dorsally (A) 2

1B. Tail not or only moderately compressed laterally, without obvious median double keel dorsally (B) 34

2A. Caudal scales arranged in regular rings, occasionally incomplete on the sides of the tail 27

2B. Caudal scales not arranged in regular rings, as ventral caudal scales are larger than dorsal caudals. 3

3A. Nostril an oblique split (C) 4

3B. Nostril oval or round (D) 11

4A. Nostril closer to orbit than to tip end of snout (E) 5

4B. Nostril nearer to tip of snout than to orbit (F) 10

5A. Nuchal scales enlarged, posterior ones keeled. 9

5B. Nuchal scales not enlarged and keeled 6

6A. Chin and throat with transverse black bars; median supraocular scales transversely enlarged 8

6B. Chin and throat with black spots; supraocular scales not enlarged 7

7A. Scales on crown larger than nuchal scales . . . . . *V. b. bengalensis*

7B. Scales on crown smaller than nuchal scales. . . . *V. b. irrawadicus*

8A. Transverse bands on tail . . . . *V. b. nebulosus*

8B. No transverse bands on tail . . . . *V. b. vietnamensis*

9A. Scale rows around middle of body more than 80 . *V. d. dumerilii*

9B. Scale rows around middle of body less than 80 *V. d. heteropholis*

10A. Ground color yellow with alternating transverse bars of reddish brown on back and tail . . . . *V. flavescens*

10B. Ground color olive with transverse black bands on back only . . . . *V. olivaceus*

11A. Nostril nearer to tip of the snout than to orbit 12

11B. Nostril nearer to orbit than to tip of snout . . . . *V. niloticus*

12A. Nostril lateral 13

12B. Nostril on top or end of snout. 20

13A. Tail less than 2× length of head and body 14

13B. Tail more than 2× length of head and body . . . . . . *V. salvadorii*

14A. Throat yellowish, whitish or spotted 15

14B. Throat light red (peach) . . . . *V. jobiensis*

15A. Yellow spots on body larger than 4mm, arranged in transverse rows. 16

15B. Yellow spots on body smaller than 2mm, not in any distinct pattern. . . . . *V. indicus*

16A. Dorsal yellow spots in four transverse rows . . . . . . *V. spinulosus*

16B. Dorsal yellow spots not in 4 rows 17

17A. Dorsal yellow spots in 10–12 transverse rows; blue reticulations on the tail. 18

17B. All supraocular scales small 19

18A. Throat dark with large white spots . . . . *V. d. doreanus*

18B. Throat mostly white . . . . *V. d. finschi*

19A. Back, sides, and legs with large, black-ringed, yellowish spots; snout broad . . . . *V. giganteus*

19B. Back with small yellowish spots; usually black bars across pointed snout; legs with yellow bands or large spots . . . . *V. varius*

20A. Nostril at tip of snout 21

20B. Nostril on top of snout; very strong caudal keel. . . . . *V. mertensi*

21A. Snout long, depressed at the end 22

21B. Snout round, blunt . . . . *V. komodoensis*

22A. Ground color dark brown to olive or yellowish, with yellow spots and transverse rows of larger ocelli 23

22B. Ground color almost black, yellow spots very faint . . . . *V. s. andamanensis*

23A. Head mostly brown. 24

23B. Head olive with yellow spots . . . . *V. s. salvator*

23C. Head bright yellow . . . . *V. s. cumingi*

24A. Ground color yellowish brown 25

24B. Ground color grayish or blackish brown 26

25A. Yellow spots on body bright . . . . *V. s. bivittatus*

25B. Yellow spots on body obscure . . . . *V. s. marmoratus*

26A. Gray-brown above with light vertebral stripe . . . . . *V. s. nuchalis*

26B. Black-brown above with a yellow spot on each scale . . . . *V. s. togianus*

27A. Tail at least 1.3× as long as head and body; scales on upper base of tail not rugose. 28

27B. Tail less than 1.2× as long as head and body; scales on upper base of tail rugose . . . . *V. spenceri*

28A. Scales on head and body small. 29

28B. Scales on head and body large 33

29A. Dorsal pattern of alternating transverse rows of small pale spots and large dark spots 30
29B. Dorsal pattern without transverse rows of large dark spots . . . 31
30A. Dorsal ground color dark brown. . . . . *V. g. gouldii*
30B. Dorsal ground color grayish brown.. . . . . *V. g. horni*
30C. Dorsal ground color reddish brown. . . . . *V. g. rubidus*
31A. Neck with transverse dark bands.. . . . . *V. rosenbergi*
31B. Neck without transverse dark bands 32
32A. Dorsal ground color reddish brown.. . . . . *V. f. flavirufus*
32B. Dorsal ground color dark brown . . . . *V. f. gouldii*
33A. Nuchal scales greatly enlarged, posterior ones strongly keeled. . .
. . . . *V. rudicollis*
33B. Nuchal scales not greatly enlarged, not keeled. . . . . . *V. mitchelli*
34A. Tail round or somewhat dorsoventrally flattened 35
34B. Last two-thirds of tail somewhat compressed laterally . . . .
. . . . *V. semiremex*
35A. Nostril an oblique slit 36
35B. Nostril oval or round 43
36A. Snout dorso-ventrally flattened at the end. 37
36B. Snout convex. 39

G

37A. Last half of tail distinctly keeled. . . . . *V. g. koniecznyi*
37B. No part of the tail distinctly keeled 38
38A. Dorsal spots yellow; banding obscure or absent in adults . . . . . .
. . . . *V. g. griseus*
38B. Dorsal spots brown; distinct banded pattern . . . . . . *V. g. caspius*
39A. Snout strongly convex (G); sides of throat pure white 40
39B. Snout slightly convex (H); sides of throat not white . . . .
. . . . *V. exanthematicus*

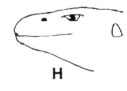

H

40A. Transverse rows of white spots on the dorsum 41
40B. No obvious dorsal white spots . . . . *V. yemenensis*
41A. Dorsal white spots with black edges. 42
41B. Dorsal white spots without dark edges; dark temporal stripe does not extend as far as the shoulder . . . .
V. a. microstictus or ***V. a. ionedesi*** (differ only in juvenile patterns)
42A. More than 140 rows of scales around middle of body . . . .
. . . . *V. a. albigularis*
42B. Less than 140 rows of scales around middle of body, scales large
. . . . *V. a. angolensis*
43B. Nostril closer to tip of snout than to eye 44
44A. Dorsal and lateral tail scales may be smooth or keeled but very .
spiny. 45

44B. Tail scales may be smooth or keeled but not spiny 47
45A. More than 70 rows of scales around middle of body 46
45B. Fewer than 70 rows of scales around middle of body . . . .
. . . . *V. primordius*
46A. More than 80 rows of scales around middle of body . . . .
. . . . *V. s. storri*
46B. Fewer than 80 rows of scales around middle of body . . . .
. . . . *V. s. ocreatus*
47A. Scales on top of head smooth 48
47B. Scales on top of head keeled. . . . . *V. eremius*
48A. Tail with transverse bars 49
48B. Tail uniform in color. 55
49A. Tail patterned throughout its length. 50
49B. Tail with transverse bars on proximal end only. 53
50A. Tail with narrow white bands; dorsal and lateral caudal scales moderately to strongly keeled 51
50B. Tail with conspicuous alternate black and white bands; caudal scales not keeled . . . . *V. glauerti*
51A. Distinct white ocelli on back; caudal scales strongly keeled. . . 52
51B. Small white or yellow spots on back; caudal scales only moderately keeled . . . . *V. t. similis*
52A. Gray cross-bands on back . . . . *V. scalaris*
52B. No cross-bands present on back . . . . *V. t. timorensis*
53A. Last half of tail black 54
53B. Last half of tail creamy white. . . . . *V. glebopalma*
54A. Head and neck mostly black . . . . *V. t. tristis*
54B. Head and neck with whitish ocelli . . . . *V. t. orientalis*
55A. Tail uniformly green 56
55B. Tail uniformly black 57
56A. Ventral color uniform. . . . . *V. prasinus*
56B. Ventral color mottled . . . . *V. telenestes*
57A. Body uniformly black. 58
57B. Ventral color yellowish . . . . *V. teriae*
58A. Nuchal scales keeled. . . . . *V. beccarii*
58B. Nuchal scales smooth, tubercular. . . . . *V. bogerti*
59A. Dorsal and lateral tail scales very spiny 60
59B. Tail scales smooth or keeled but not spiny. 63
60A. Dorsal pattern of ocelli 61
60B. No dorsal ocelli. . . . . *V. baritji*
61A. Distinct pattern of ocelli present. 62

61B. Pattern of ocelli broken, irregular; very dark ground color . . . . .
. . . . *V. a. insulanicus*
62A. Dark brown stripe through eye . . . . *V. a. acanthurus*
62B. Yellow stripe through the eye . . . . *V. a. brachyurus*
63A. Tail longer than head and body 64
63B. Tail shorter than head and body. . . . . *V. brevicauda*
64A. Tail with longitudinal stripes 65
64B. Tail without longitudinal stripes. 66
65A. Body and base of tail with narrow purplish-brown cross-bands
. . . . *V. gilleni*
65B. Body and base of tail with scattered dark brown spots . . . .
. . . . *V. caudolineatus*
66A. Several rows of enlarged keeled scales on each side behind the vent . . . . *V. pilbarensis*
66B. No enlarged keeled scales behind the vent . . . . *V. kingorum*

Chapter 3
ANATOMICAL AND PHYSIOLOGICAL SPECIALIZATIONS OF VARANIDS

VOMERONASAL SYSTEM AND OLFACTION

The sense of smell is very important to monitors in detecting food. Each external naris (nostril) opens into a nasal capsule. This is a narrow, elongated, U-shaped tube which leads posteriorly to an opening through the palate into the mouth. These paired openings are the internal nares (choanae) (Fig. 3.1). Several chambers exist inside the nasal capsule, one of which is for olfaction (smell). The nasal capsule is composed of cartilage and is covered anteriorly only by skin. The external naris opens to the vestibulum, a nonsensory chamber of the nasal cavity. Cavernous erectile tissue surrounds the beginning of the vestibulum and may act as a valve to close the external naris. The second chamber is the olfactory chamber proper, and is lined with a columnar sensory epithelium with bipolar neurons. This olfactory membrane is not as extensive in varanids as it is in other lizards.

The duct of the external nasal gland also enters the vestibulum. The gland itself lies lateral to and outside the nasal capsule. It functions in the excretion of excess sodium and will be discussed further later in this chapter.

All varanids with slit-shaped external nares are able to close the slit completely by pushing the inner medial wall of the naris against the outer lateral wall. This may help keep debris out of the nose while the snout is used to poke around leaves and other ground debris during food search.

In addition to the olfactory chambers, varanids have a second pair of olfactory organs called the vomeronasal or Jacobson's organs. The vomeronasal organs are much larger in varanids than in other lizards. They are located in the cartilage of the nasal capsule, anterior to the choanae. They open via a narrow duct into the roof of the mouth. The vomeronasal organ also contains bipolar neurons which form a branch of the olfactory nerve which leads to the accessory olfactory bulb of the brain.

Behavioral studies indicate that the olfactory chamber membrane can alert a resting monitor to the proximity of food, although it is possible that airborne molecules might pass through the choanae, into the mouth, and thence into the vomeronasal organ. Hunting monitors, however, rely heavily on the vomeronasal organs. Using the tongue to pick up molecules from the environment, these molecules are deposited on paired pads on the floor of the mouth, the sublingual plicae, which are then pressed against the duct openings in the roof of the mouth. This is done continuously when the lizard is actively hunting.

The sense of smell is highly developed. Komodo monitors (*V. komodoensis*) are able to detect the scent of carrion from as far as 11 kilometers (Auffenberg, 1981a). Bengal monitors (*V. bengalensis*) have been tracked walking three kilometers attracted by the scent of carrion (Auffenberg, 1994). Terrestrial species can detect food under several inches of soil and dig it up.

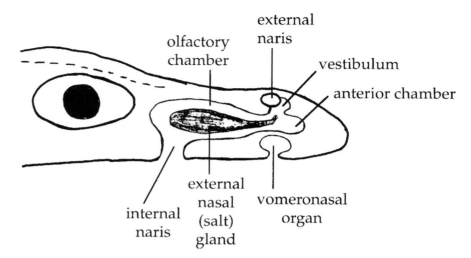

Figure 3.1. Generalized anatomy of the nose region of *Varanus*.

Tsellarius and Men'shikov (1994) report that *V. griseus* can distinguish the age and sex of conspecifics from their scent trails for 2–3 days. Desert monitors mark their trail by dragging the cloaca on the ground for up to 10 m. This behavior occurs around shelters, and when crossing the fresh trails of conspecifics. The behavior declines with age among males, as they rise higher in the local hierarchy, and it occurs less frequently in females than males. During the mating season, when a male comes across the trail of a female, he will turn and follow it. If the trail is fresh he invariably heads in the direction taken by the female, but if the trail is old he may sometimes be unable to distinguish the direction the trail leads.

VISION

By reptilian standards monitors have excellent vision. Horn and Visser (1989) claim that *V. giganteus* can spot an upright human from 300 m; Stanner (1983) actually measured 200 m vision range in *V. griseus*. Auffenberg (1994) measured a 250 m range for *V. bengalensis*. Although the eyes are placed laterally, the pointed snout of most species gives limited (25% overlap) stereopsis (depth perception) (Fig. 3.2). This is a valuable adaptation for a predator aiming a strike. The visual field encompasses about 240°.

The eye is covered by two unequal lids. The upper lid has little mobility. The lower lid contains a cartilaginous plate, the tarsus, which slides over the surface of the eye. The eye is opened mostly by contracting the lower lid. On the anterior side of the eye is a fold of conjunctiva called the nictitating membrane. The nictitating membrane is drawn across the cornea by muscular action and contains folds for catching particles of dirt. Some general features of the eye are illustrated in Fig. 3.3.

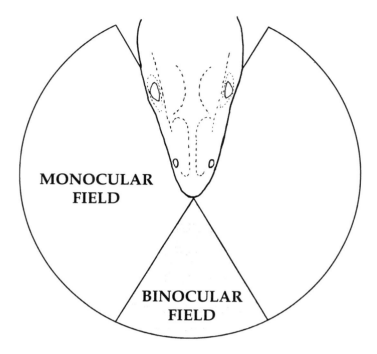

Figure 3.2. Field of vision for a typical varanid.

The lens of the varanid eye is fairly soft and accommodation for near vision is accomplished by squeezing of the lens by the ciliary muscles to make it more spherical. The iris forms a round pupil.

The scleral tunic is partly fibrous and partly cartilaginous. Fifteen scleral ossicles form a ring covering the exposed part of the eyeball (Fig. 3.4). Thus, the eyeball of monitors is more heavily protected than the human eye.

The retinal tunic is similar to that of most vertebrates. It has a central fovea but possesses only cone photoreceptors. It is concluded that monitors have excellent color vision, but see poorly or not at all in dim light. A unique device for increasing retinal nutrition is an elongated cone of tissue which projects into the semifluid vitreous from the back of the eye. It is thought to secrete nutrients for the neurons of the inner retina.

Vision plays the primary sensory role in a monitor's pursuit of live prey and in avoiding predators.

THE EAR AND HEARING

The varanid ear really has few specializations over the "typical" lizard ear, but this pair of important senses is still worth considering briefly. As in humans, the ear is a complex organ which functions both for maintaining balance and for the reception of sound. In reptiles, the former function is believed to be

THE NATURAL HISTORY OF MONITOR LIZARDS

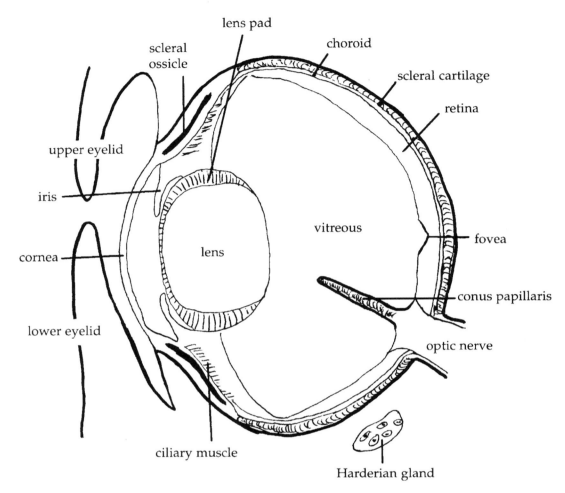

Figure 3.3. Anatomy of the major features of the eye of *Varanus*.

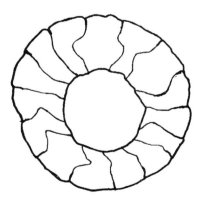

Figure 3.4. Scleral ossicles of the eye of *Varanus*.

primary. The ear is usually divided into three regions—the external ear, middle ear, and inner ear.

The external ear of lizards, including varanids, consists only of a depression on the posterior side of the head. The size and shape of the external ear opening varies considerably (Fig. 3.5) between species, and is presumably adaptive to different life strategies (Auffenberg, 1994). At the bottom of the depression is the tympanic membrane (eardrum). It is thin and covered externally by modified skin.

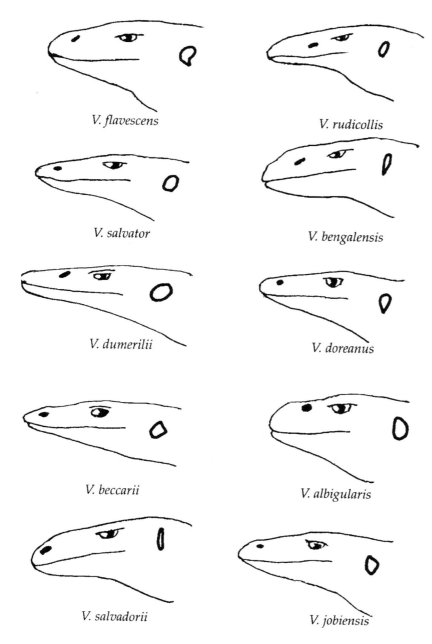

Figure 3.5. External ear openings of various species of varanids.

Beneath the tympanic membrane is a space in the skull, the tympanic cavity, which houses the structures of the middle ear. The tympanic cavity is connected to the pharynx by the auditory (Eustachian) tube which helps equalize the pressure inside the cavity with that in the throat and thus that of the outside. The tympanic cavity is lined with a mucous membrane.

Chief among the structures of the middle ear of reptiles is a single bone, unlike the three bones in our own middle ear. This bone, the stapes or columnella, conducts vibrations from the tympanic membrane, across the tympanic cavity, to the inner ear. The bone is rod shaped with a cartilage attachment (extrastapes) to the tympanic membrane at one end and an expanded footplate at its inner end (Fig. 3.6). The stapes also articulates with the quadrate

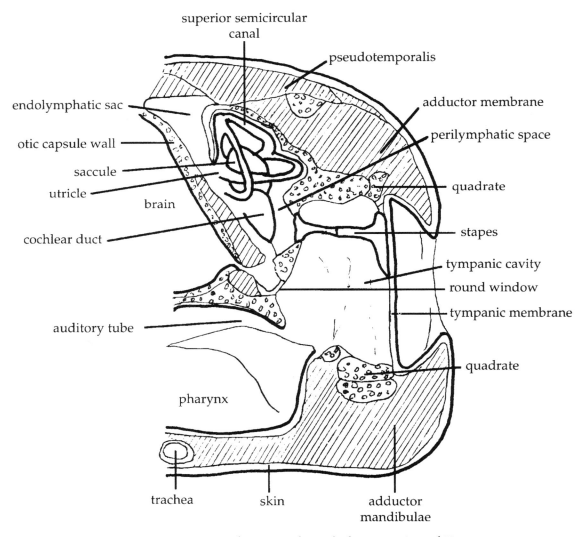

Figure 3.6. Cross (coronal) section through the ear region of *Varanus*.

bone, one of the bones of the jaw which probably became, in the course of evolution, one of the other bones (malleus) found in the ear of mammals.

The inner ear consists of a bony otic capsule inside of which is an intricate series of membranous tubes and sacs. This arrangement is common to all vertebrates. The upper part of the otic capsule contains the three semicircular canals and ducts, two vertical ones and a horizontal one, all at right angles with each other and connected to a round chamber, the utricle. The utricle in turn is connected to another chamber of about equal size in varanids, the saccule. The saccule then connects with the cochlear duct which contains the receptors for hearing sound.

The otic capsule contains a fluid called perilymph. The six membranous sacs and tubes, described above, are filled with another fluid, the endolymph. It is within these sacs and tubes that are found the patches of sensory cells for hearing and equilibrium.

Little experimental work has been done with the equilibrium sense in reptiles. The semicircular ducts each have an ampullary crest of sensory cells bearing a tuft of stereocilia on each, giving them the name hair cells. The inertia of the endolymph when the head is turned in any direction causes the stereocilia to be bent, setting up nerve impulses in the connecting nerve fibers. This is called the dynamic equilibrium sense.

The utricle and saccule also contain patches of hair cells (the maculae) which give information about static posture, sensing the tilt of the head, and about acceleration, as when a monitor makes a sudden lunge at prey. It is worthy of note that these five parts (the so-called vestibular apparatus) have remained largely unchanged throughout vertebrate evolution, keeping us right side up as effectively on land and in the air as in the water.

Joined to the posterior end of the saccule is the cochlear duct, shaped somewhat like a pyramid. The lizard cochlea is more rudimentary than that of humans in whom it has become longer and coiled. The cochlea of varanids is the most complex among lizards, approaching the three-chambered form found in mammals. Its principal sensory area is an elongated strip of tissue called the papilla basilaris, which serves the same function as the more elaborate organ of Corti in the mammalian cochlea. The papilla basilaris sits on resonating fibers, the basilar membrane, which respond differentially to various vibration frequencies. The papilla basilaris contains the actual receptor cells for sound, also called hair cells. There are about 1,200 hair cells in *V. bengalensis*, 900 in *V. niloticus*, and 400 in *V. griseus* (Wever, 1978). These numbers can be compared to 300 in *Iguana iguana* (Common Iguana) and 1,600 in *Gekko gecko* (Tokay Gecko). Geckos do use sound for communication. The number of hair cells is believed to be correlated to sensitivity to sound. The 23,500 hair cells in the human organ of Corti probably reflect the far greater importance of sound in our lives than in the lives of lizards. These hair cells are stimulated by sound vibrations picked up by the tympanic membrane and transmitted by the stapes to the perilymph and endolymph and thence to the basilar membrane. The papilla basilaris responds to frequencies between 100 and 10,000 Hz., with peak sensitivity in the 400-2,000 Hz range for varanids (Wever, 1978). Humans respond to frequencies between 20 and 20,000 Hz, and of course we by no means possess the broadest range of hearing among mammals. Behavioral experiments indicate not only a falling off of sensitivity in the varanid cochlea for frequencies above 2,000 Hz but also an insensitivity to low decibel (sound intensity) levels that humans can detect.

EPIDERMAL SENSES

Mammals, such as humans, have a large number of specialized sensory receptors in the skin, subserving such functions as touch, pressure, heat, cold and pain detection. Reptiles lack most of these skin senses. Varanids do possess raised sensory papillae on their scales. At least one papilla is found on almost every scale of the body and limbs (Fig. 3.7). A sensory papilla consists of a group of columnar cells with invaginated free nerve endings about 100 micrometers in diameter. These sensory papillae are believed to function as a fine sense of touch, but this has not actually been demonstrated so far.

Lamellated receptors resembling the Pacinian corpuscles of mammals are found on such areas as the foot pads. These are surrounded by a capsule and can be as large as 0.5 mm. These are probably rapidly adapting tactile organs (von During and Miller, 1979) involved with vibratory sense and touch-pressure sense.

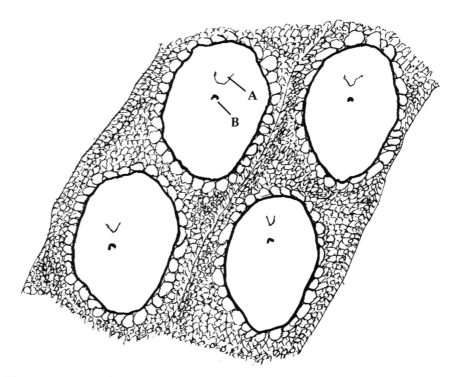

Figure 3.7. Typical dorsal scale pattern of *V. albigularis*, showing large central scale and surrounding rosette of small scales. A, opening to exocrine skin gland; B, raised sensory papilla.

TONGUE AND HYOID APPARATUS

The original broad, flat form of a lizard tongue is primarily for handling food and only secondarily for conveying odor molecules to the vomeronasal organs. In varanids, however, the latter role has superseded the former almost entirely. This gives monitors a superficial resemblance to snakes. Since the musculature of the monitor tongue differs from that of snakes, the forked tongue is believed to be an example of convergent evolution. The varanid tongue lacks taste buds (Schwenk, 1985), although they are abundant in other lizards which have the usual rounded tongue.

The varanid tongue is long and narrow (Fig. 3.8). It is deeply divided at the tip and fits into a sheath when retracted. It lacks the rough dorsal surface of most lizard tongues, and consists of four parts: (1) a posterior part with longitudinal muscles; (2) a short section with chiefly circular muscles; (3) a body with very complexly arranged muscles; and (4) the forked tip.

The tongue is extended and flicked rather rapidly. It is moved up and down through an arc sampling a field of air of about 7 cm^2. Particles are then carried back to the mouth and thence to the duct of the vomeronasal organ. The forked tines (halves) of the tongue are spread wide apart at the end of each extrusion. This enables the lizard to pick up gradients instantaneously in a chemical trail and thus follow the trail efficiently (Schwenk, 1994). Monitors are able to increase the tongue length 70–90 percent when it is fully extended. The flicking tongue is characteristic of monitors as they search for food or explore new areas.

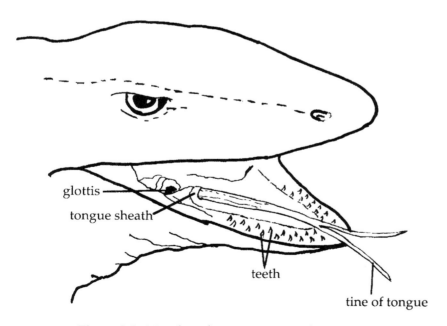

Figure 3.8. Mouth and tongue region of *Varanus*.

THE NATURAL HISTORY OF MONITOR LIZARDS

The tongue is partially attached to a complex skeletal structure called the hyoid apparatus (Fig. 3.9). All snakes and lizards have such an apparatus which is helpful for swallowing large food items. That of varanids differs somewhat from other lizards because its joints are more flexible and the hyoid process enters the connective tissue below the tongue. The apparatus is larger than in other comparably sized lizards.

Swallowing is accomplished by the broad throat muscles, flexible jaws, and hyoid apparatus. Prey is pushed backwards in the mouth, and the jaws then move forward to engulf it. Two or three of these pushing and engulfing

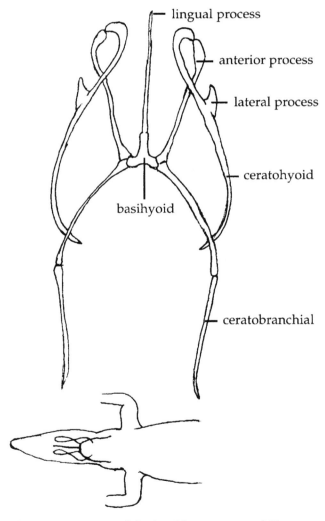

Figure 3.9. Bones of the hyoid apparatus of *Varanus*.

movements are usually sufficient for small prey, but larger items may require several more. Monitors often use the ground or other objects in the environment (such as cage walls) to help manipulate large food items. The food is pushed against the object to help force it down the throat. The tongue is withdrawn into its sheath during ingestion. Monitors have been observed to use it, however, to clean pieces of food from their teeth, and in captivity at least, it is used to lick the last juice from a food dish. Monitors have not lost the eating function of the tongue as completely as snakes have.

Once food has reached the gular region of the pharynx, the hyoid apparatus moves it back to the esophagus. The neck muscles are also brought into play by bending from side to side to help move food items down the esophagus, much as snakes do.

The hyoid apparatus is also used in drinking. The snout is immersed and water is pumped into the throat by the hyoid apparatus. The head is then lifted and the throat compressed in swallowing.

The hyoid apparatus is also used to extend the gular region in threat display behavior.

All monitors have two pairs of salivary glands associated with the lower jaw. The infralabial gland is the superior of the two and begins near the jaw tip in the floor of the mouth. The mandibular gland is located beneath the infralabial and has a single duct opening near the base of the fifth dentary tooth. It is the mandibular gland that has become the venom gland in the Gila monster and beaded lizard (*Heloderma* spp.). There is no credible evidence of venom being produced in any varanid in spite of it being a frequent belief in many rural areas of Asia.

TEETH AND JAWS

The role of varanids as carnivores is keenly evident in their dental specializations. The teeth are typically large and sharp. They are compressed laterally, curved posteriorly, with a sharp tip and broad base (Fig. 3.10). Curved teeth are better for catching and holding prey than straight teeth. The teeth are not as easily seen as in crocodilians because they are hidden within a fold in the gums. However, they become readily evident when the monitor bites. I have personal acquaintances who have visited the emergency room for stitches due to deep lacerations from bites of meter-length monitors.

The teeth of the larger species (e.g., *V. komodoensis*) are serrated on the posterior edge allowing them to cut the flesh off large prey and carrion. Some species (e.g., *V. olivaceus* and *V. niloticus*) are exceptions to the typical tooth morphology. Adults of these species have blunt, peg-like teeth, reflecting the importance of snails in their diet. In most monitors there are 7–9 premaxillary, 10–13 maxillary, and 12–13 mandibular teeth. The number is usually constant for one species.

Monitor teeth are termed pleurodont, which means that they are attached to the side of the jaw. In this position they are easily broken, but most species have 2–3 rows of replacement teeth which develop from the pulp

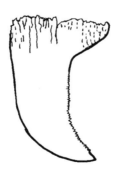

Figure 3.10. Maxillary tooth of *Varanus komodoensis*, after Auffenberg (1981a).

cavity and become fused to the lateral side of the jaw when the old tooth is gone. In *V. bengalensis* the average functional life of a tooth is about three months (Edmund, 1969), with tooth replacement averaging four times yearly. Some species apparently have even more rapid turnover.

The muscles of the jaws and throat of all but the smallest monitor species are adapted for feeding on large prey, even if they only infrequently do so. All species' jaws close rapidly enabling them to capture fast-moving prey. The prey is usually bitten several times and then pressed against objects in the process of being killed. Larger prey are often smashed against the ground or branches. Prey are held by the jaws until all movement ceases. Small prey are swallowed whole, usually head first. The larger species of monitors are able to tear prey items into pieces. They hold the prey down with their front feet and tear away pieces of flesh with powerful neck muscles. A 50 kg Komodo monitor can consume a 37 kg wild boar in 17 minutes this way (Auffenberg, 1981a).

DIGESTIVE TRACT

The morphology of the digestive tract is highly correlated with diet in most vertebrates. This is also true of monitors. The intestines of carnivores such as varanids tend to be short because their diet is high in proteins and lipids that do not require a great deal of digestive processing. These species may also have quite large stomachs to accommodate large amounts of food when it is available, as in the case of the Komodo monitor mentioned above. Only *V. olivaceus* among monitors has the gut of a herbivore—long intestines, and an especially large colon. Table 3.1 summarizes the volumetric relations of parts of the digestive tract of several varanid species, corrected for body length.

TABLE 3.1
Volumetric Relations of Parts of the Digestive Tract of Varanid Lizards (modified from Auffenberg, 1994) (in cm^3)

Species	Stomach	Small intestine	Large intestine
V. bengalensis	57.9	42.2	20.0
V. flavescens	69.2	43.6	38.6
V. griseus	106.0	26.2	19.8
V. olivaceus	46.9	52.8	64.4
V. rudicollis	118.8	94.3	46.6
V. salvator	138.8	34.0	5.6

This data indicates that *V. salvator* is the most highly developed (on this table) as a carnivore.

The liver is a large organ in all varanids, constituting more than 2 percent of body weight (Fig. 3.11). There is considerable variance among species, and between juveniles and adults. The differences are believed to be largely due to differences in diet. Liver function is believed to be similar to that of mammals. Glycogen storage, cholesterol synthesis, blood protein synthesis, bile synthesis, and detoxification of many substances are probably the most important functions. The liver is relished by rural people of eastern India and Burma.

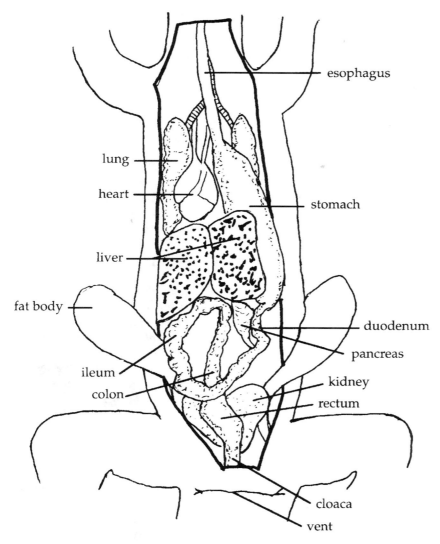

Figure 3.11. Internal anatomy of a varanid lizard.

Digestive efficiency is high in varanids ranging from 70 to 90 percent. Hair, feathers, claws, teeth, chitin, and shells of molluscs and eggs are about the only parts of prey that are not digested. Digestion is slowed or stopped in lizards that are under stress, and partially digested prey may be defecated.

Digestive time is largely a function of temperature. At near preferred body temperatures (30° to 35°C) passage time averages about 26 hours. In the wild, lower night temperatures will often prolong this time. The mean time required for passage in free-ranging *V. komodoensis* is 4.7 days.

Varanids cast gastric pellets consisting of tightly packed undigestible parts, mentioned above, in a foul-smelling mucus-covered mass similar to those of owls. They are disgorged by a regurgitative action. The body and neck are arched up off the ground, and the neck is thrown from side to side as the stomach and esophageal muscles go into reverse peristalsis.

LOCOMOTION

Terrestrial

All monitors are quadrupedal in terrestrial locomotion. In walking the quadrupedal gait is similar to other lizards with the lateral undulations of body, head, and tail more exaggerated by the longer body form. The two diagonally opposite feet are raised and advanced while the other two support the weight of the lizard and propel it forward. The head normally sways to one side as the tail swings to the opposite side. Normal walking speed ranges from 0.5km/hr in smaller species to 4.8km/hr for *V. komodoensis* (Auffenberg, 1981a). All previously studied terrestrial species spend a great deal of time walking.

In running, the degree of lateral undulation is reduced as the body and tail are held somewhat rigid. The hind feet are moved in a wide lateral arc. The larger terrestrial monitors can attain 15km/hr over short distances. *Varanus bengalensis* has a mean running speed of 6.8 km/hr, and a maximum of 17.2 km/hr (Auffenberg, 1994). A few of the larger species, after gaining acceleration running on four feet, are reported to sometimes shift to bipedal locomotion. What advantage they might gain by this is unknown.

Climbing

All monitors climb well, and some go as high as 30m or more in the trees. A few species, such as *V. prasinus* and *V. gilleni*, are highly adapted for an arboreal life, having a prehensile tail. Others, such as *V. glebopalma*, are very adept at climbing vertical rock faces. Even desert forms, such as *V. albigularis* from Namibia, often spend considerable time in the tallest tree or shrub in their home range (Phillips and Alberts, 1994). Juveniles of terrestrial species are typically more arboreal than adults.

TABLE 3.2 Degree of Arboreality in *Varanus* Species

Class	Representative species
1. Climb little	*V. griseus*
2. Known to spend time in trees but without specialized structures for climbing	*V. komodoensis, V. salvator, V. albigularis, V. bengalensis, V. exanthematicus*
3. Scansorial, spending much time in trees, with arboreal specializations	*V. olivaceus, V. rudicollis, V. gilleni, V. salvadori*
4. Arboreal, highly specialized for climbing, little terrestrial foraging	*V. prasinus, V. beccarii*
5. Extreme adaptation for arboreal life, no terrestrial activity	None

Swimming

Almost all monitor species are good swimmers. Some species are very aquatic, e.g., *V. indicus, V. salvator, V. niloticus, V. mertensi*. Swimming is apparently accomplished by the lateral undulatory movements used for walking. Underwater photography reveals that *V. salvator* folds its legs back against the body to reduce drag and swims by undulating the body. The tail is not used by the water monitor, although Auffenberg (1981a) claims that the Komodo monitor uses tail undulation as well.

There are reports of *V. salvator* swimming several kilometers from one island to another (Smith, 1932; Annandale *in* Boulenger, 1903). However, no one has actually tracked a monitor through the water. Auffenberg (1981a) records *V. komodoensis* swimming regularly across a 450m tidal channel. *V. niloticus* can remain underwater for as long as one hour. *V. salvator* has been recorded diving to at least 8m. What monitors due at such depths is unknown, as they have never been seen foraging underwater. In fact, underwater photos of *V. salvator* show that the eyes are closed when it is underwater. *V. mertensi*, however, has been seen *walking* underwater apparently searching for crabs or fish (King and Green, 1993). The escape behavior of many species includes diving and swimming away underwater. There is also speculation that some tropical species may spend time underwater to cool off. Even desert species such as *V. griseus* and *V. varius* swim well and do not hesitate to enter water. The highly terrestrial *V. bengalensis* will dive underwater to escape pursuit if the opportunity arises, and can remain submerged for at least 17 minutes (Auffenberg, 1994).

THE NATURAL HISTORY OF MONITOR LIZARDS

RESPIRATION

The millions of cells of the vertebrate body require a constant supply of energy. This energy is needed to help cells perform their many chemical activities in maintaining the life of the animal. Oxygen must be in constant supply to facilitate the release of energy stored in nutrient molecules. The exchange of oxygen and carbon dioxide between the body and the outside environment is known as external respiration. In terrestrial vertebrates it is performed by the lungs.

Varanids have several specializations in which their lungs differ from those of other reptiles. Typical reptiles do not possess a diaphragm as do humans and other mammals. The lungs are contained in a single ventral cavity with the abdominal organs. Varanids, however, have a septum which separates the abdominal cavity from the cavity containing the heart and lungs (Figs. 3.11 and 3.12).

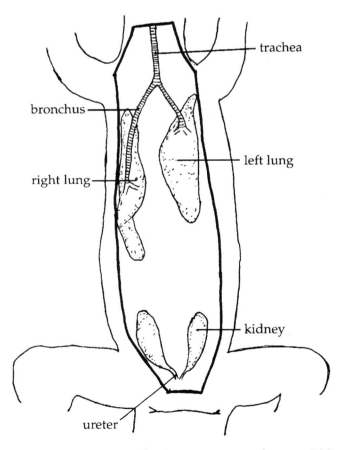

Figure 3.12. Respiratory and urinary anatomy of a varanid lizard.

In mammals the diaphragm is the chief breathing muscle. Lacking this muscle, typical reptiles use the intercostal muscles to draw the ribs upward and outward increasing the size of the pleural cavity. This creates a lower than atmospheric pressure in the lungs and air flows into the lungs. When these muscles relax, the collapse of the cavity forces the air out. Reptile lungs typically have six times the volume that similar sized mammals have, but their simpler internal structure yields less surface area, which is the key element in lung efficiency. However, varanid lungs have a larger volume than most reptiles, enabling them to take in relatively large amounts of oxygen. They also have greater internal surface area than other lizards, making them more efficient (Fig. 3.13). Varanid lungs are attached to the rib cage preventing them from collapsing completely during normal breathing. They have a strong compliance with body movements and can thus be very efficiently ventilated by this costal breathing.

Reptile lungs must function over a wide range of oxygen needs depending on the activity levels of the animals. These needs are always less than similar-sized birds or mammals whose energy needs are much greater (the high cost of being "warm-blooded"). The trachea extends from the glottis in the mouth to the pleural cavity where it divides into two bronchi, one to each lung. The walls of the trachea and bronchi are reinforced with cartilage to keep them from collapsing. Each bronchus continues into the lung where it connects with many small air sacs. This is called a multicameral lung. The surface area, to which gas exchange is directly related, is increased further by the presence of smaller chambers called faveoli. These are functionally equivalent to the alveoli of human lungs, but 100 times larger. With enhanced gas exchange, varanids are able to maintain higher levels of aerobic activity than other lizards. The respiratory membrane of varanids is able to absorb oxygen at levels equal to that of mammals at rest. This gives monitors a high aerobic capacity—the ability to support activity using oxygen-based metabolism.

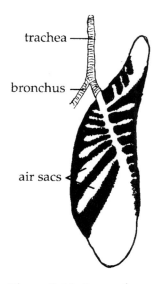

Figure 3.13. Internal structure of the varanid lung, after Becker et al. (1989).

Most reptiles rely heavily on anaerobic metabolism to support the high activity required for running, subduing prey, etc. Anaerobic respiration does not use oxygen to release energy from nutrient molecules, but it is much less efficient as most of the energy remains trapped in the form of lactic acid. Anaerobic metabolism has the advantage of working under less than optimal body temperatures. It is of great advantage to a reptile whose body has not yet reached the temperature required for aerobic respiration to begin. It has the disadvantage of building up large quantities of lactic acid which quickly causes fatigue by accumulating in the muscles. Lactic acid is removed as the reptile returns to rest and aerobic metabolism. Lactic acid is converted aerobically into carbon dioxide and water. With their greater aerobic scope varanids are able to maintain higher levels of activity for longer periods of time than other lizards. However, recent field research with radio telemetry (Christian and Weavers, 1994) indicates that some species of varanids (*V. flavirufus*, *V. rosenbergi*) actually spend less time each day in activity than large terrestrial iguanas (i.e., *Conolophis pallidus* [Galapagos land iguana], *Cyclura nubila* [Cuban iguana]).

Using a treadmill test Bickler and Anderson (1986) found that the mostly arboreal *V. gilleni* could run for many minutes at 1.0 km/hr with no signs of fatigue. A similar sized terrestrial lizard, the desert iguana (*Dipsosaurus dorsalis*) could only reach 0.8 km/hr before anaerobic metabolism set in. It seems probable that the higher aerobic capacity of varanids is related to the intensity of their activity and would be advantageous in capturing moving prey.

In most lizards blood is shunted away from the lungs during periods of inactivity and breathing becomes irregular. Varanid lungs, however, receive a continuous flow of blood. Breathing is regular and low during inactivity. *V. albigularis* breathes 7–8/min at 35°C, awake, and 3/min asleep at 25°C (De Lisle, unpubl.)

CIRCULATION

In order for the cells of a varanid to make use of the more efficient external respiratory system, monitors require a circulatory system more efficient than that of other lizards.

The varanid heart is not a typical lizard heart. It is located more posteriorly than most lizards, right behind the posterior end of the sternum, between the lobes of the liver. It is surrounded by a thick pericardium. The internal structure (Fig. 3.14) is also strikingly different from that of other lizards. The ventricular chambers are still connected as they are in other lizards, but the arrangement of the chambers and valves together with a flexible septum minimizes the mixing of oxygenated and deoxygenated blood.

During systole (heart contraction) the cavum pulmonale functions isolated from the parts of the heart involved in systemic circulation. It generates a systolic blood pressure of only 50–60 mmHg compared with 110–120 mmHg in the cavum arteriosum and cavum venosum. This is achieved by the flexible muscular septum pressing against the outer heart wall and effectively creating a four-chambered heart. The blood pressure differences between the pulmonary and systemic circulation compare favorably with those of mammals with an anatomically four-chambered heart. From this arrangement alone, the systemic circulation of varanids probably receives blood with 20 percent more oxygen than that of other lizards.

Heart rate is related to body temperature and body weight. The higher the body temperature, the higher the heart rate; the higher the body weight, the lower the heart rate. Heart rate taken for several varanids at various body temperatures ranged from 10 to 100 beats per minute (Bartholomew and Tucker, 1964).

The blood buffering system of varanids is also more effective than in other lizards. Anaerobic activity increases acid levels in the blood, especially at preferred activity temperatures. High acidity of the blood reduces the oxygen carrying capacity of hemoglobin. This in turn increases the lizard's reliance on anaerobic metabolism. A positive feedback sets in quickly leading to fatigue. Varanid homeostasis closely monitors the acidity of the blood. As body temperature rises, ventilation increases thereby eliminating excess car-

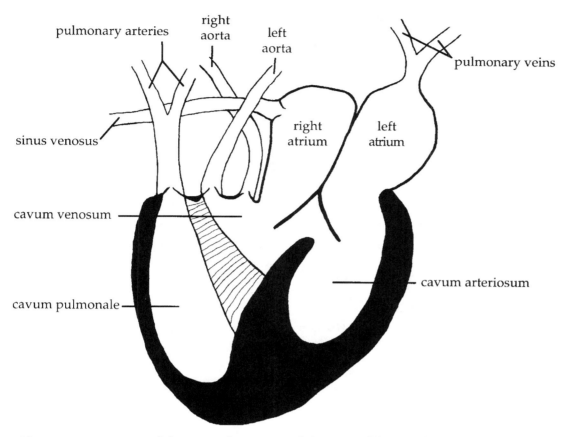

Figure 3.14. Diagram of the internal structure of the varanid heart, after Burggren (1987).

bon dioxide. This helps maintain the buffering system of the blood. Lactic acid production in varanids is thus kept at low levels allowing them to achieve high levels of activity.

Varanids also have higher levels of a muscle protein called myoglobin than do other lizards (King and Green, 1993). Myoglobin is an oxygen-storing protein; the more myoglobin present the greater sustained activity a muscle can produce.

Those varanids which are adapted to aquatic environments (e.g., *V. salvator*) appear to be more tolerant of high lactic acid levels. This tolerance allows them to stay submerged longer without coming to the surface to breathe.

The more complex structure of the heart and blood chemistry, together with the more efficient respiratory system, give varanids the ability to achieve intense activity without becoming exhausted. They have thus been able to adopt an active, wide-foraging feeding strategy. *V. salvator* has been recorded actively foraging for 13 hours a day (Traeholt, 1994).

WATER AND SALT BALANCE

One of the most important facets of an animal's internal environment is the composition of its body fluids. These fluids consist mainly of water and various dissolved electrolytes, especially sodium chloride. Water is the chief constituent of all living things, and makes up some 70 percent of the body weight of lizards (King and Green, 1993), which is about 10 percent higher than humans.

Little is yet known of the comparative water and electrolyte homeostasis in varanids. Only a few species have been studied in the wild and all of them from arid or semiarid habitats (King and Green, 1993). Many species of varanids spend considerable time in aquatic habitats or live in humid forest areas. Most of these species are known to drink water frequently. Reptiles are, however, as a group somewhat preadapted to living in arid environments by their method of excretion which requires very little water.

Terrestrial animals lose water to the environment in several ways. These include evaporation from the lungs while breathing, evaporation from the skin, and loss of water in excretion and defecation.

Evaporative water loss by varanids depends on the air temperature and humidity as well as activity level. The skin of varanids is covered by scales and contains no sweat glands. Water loss through the skin is thus very low compared to other terrestrial vertebrates. *Varanus f. flavirufus*, an Australian desert monitor, loses only 0.10 $mgH_2O/cm^2/hr$ (King and Green, 1993) as compared to 0.95 for a desert quail, and 1.48 for a nonperspiring human, at 30°C. About 70 percent of the water loss in inactive monitors is through the skin. Water loss through respiration in *V. rosenbergi* is 0.3g/hr at normal activity and temperature levels. Humans lose about 15g/hr.

Water loss rises sharply as temperature rises. It is also relatively greater in a small varanid than a large one because of the larger surface area to body weight ratio. Water loss from the surface of the eyes is also significant, about 25mg/hr in an adult *V. rosenbergi*. This may be the reason that varanids almost always bask with their eyes shut.

Water loss can be reduced behaviorally. Even in the desert, humidity inside burrows is 80–85 percent. Many species of varanids, even adult *V. komodoensis*, spend inactive periods inside burrows. This behavior may have more to do with reducing water loss than avoiding predators as evaporative water loss in a burrow is reduced almost to zero (Green, 1973). It is also noteworthy that in captivity many species (e.g., *V. komodoensis, jobiensis, rudicollis, prasinus, beccarii, indicus*) will spend the night in their water container as deep burrows are not usually available.

As a result of protein metabolism vertebrate animals produce several nitrogenous waste products. Prominent among these are ammonia, urea, and uric acid. In reptiles the chief excretory wastes are uric acid and urate salts. These are insoluble and require very little water to excrete. The reptile kidney resembles that of mammals (humans) in general morphology (Fig. 3.15), but the functional units, the nephrons, are simpler in structure. There are no loops of Henle and no osmotic gradient for concentrating urine.

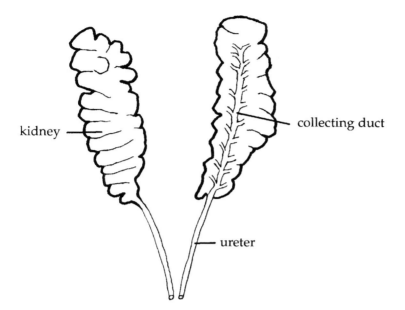

Figure 3.15. Diagram of the varanid kidneys.

In lizards the renal artery brings blood to the kidney where it subdivides several times finally entering small arterioles that service specialized capillary beds called glomeruli. Here, the blood pressure forces most of the water, salts, and other small molecules from the blood into the capsule of the nephron. This process is called glomerular filtration and the resulting fluid is the filtrate.

The filtrate first passes through a convoluted tubule where the sodium, nutrient molecules, and as much as 80 percent of the water, are reabsorbed into the blood. The remaining nitrogenous wastes, together with some additional secreted potassium and urates, are passed to the collecting ducts and finally out of the kidney to the ureter. The fluid is now called urine. The amount of urine produced depends on the water and salt balance in the animal's body. The kidneys of *V. rosenbergi* produce about 70 ml of filtrate and 3 ml of urine per hour at 30°C.

The relative number of nephrons in the reptile kidney is less than the mammalian kidney, and not all of these are operational at one time. Monitors suffering from dehydration or excessive salt loads reduce the number of operational nephrons. This is largely under control of an antidiuretic hormone, arginine vasotocin, the reptilian homolog of mammalian (human) vasopressin (ADH).

In terrestrial vertebrates, the urine in the ureters is not immediately voided, but is conducted to storage organs such as urinary bladders (in mammals) or cloacas (birds and reptiles). In reptiles the cloaca plays an important role in modifying the urine. In lizards the cloaca is divided into three chambers: (1) the coprodaeum, into which the rectum empties; (2) the urodaeum, into which the ureters empty; and (3) the proctodaeum which opens to the outside (Fig. 3.16). Urine is passed through the ureteral openings in the dorsal wall of the urodaeum and then anteriorly into the coprodaeum, which plays an important role in urine modification.

The surface area of the coprodaeum is increased by villi similar to those found in the intestine. Specialized cells in the cloacal epithelium acidify the urine as it enters the coprodaeum. This causes the resulting uric acid and urate salts to precipitate releasing considerable water which is then reabsorbed. The cloaca of *V. rosenbergi* normally reabsorbs about 8ml/H_2O/hr (King and Green, 1993). The remaining solid precipitate forms a mass inside the coprodaeum. By the time this chalky white solid mass is voided from the proctodaeum the water content has been reduced to less than half that of the most concentrated mammalian urine known.

The feces are formed in the rectum. This region of the intestine has a wall structure similar to that of the cloaca, and it reabsorbs water from the fecal material. However, the varanid rectum is not nearly as efficient in reabsorbing water as that of many other lizards adapted to arid environments. Varanid feces is still about 75 percent water.

Reabsorption of water in both the nephron tubule and in the coprodaeum is partially dependent upon active transport reabsorption of sodium from the urine. In those species which live around brackish water or even seawater, or consume invertebrate prey with high sodium content this could lead to a sodium over-load in the body. Excess sodium is removed by special salt-secreting glands found in the nasal capsules (Fig. 3.1). These glands are not unique to varanids; many lizards have them, especially herbivores like the chuckwalla (*Sauromalus* spp.) and marine iguana (*Amblyrhynchus cristatus*). Salt glands have been identified in *V. acanthurus*, *V. flavirufus*, *V. salvator*, *V. griseus*, *V. flavescens*, *V. semiremex*, and *V. rosenbergi*. However, the terrestrial *V. giganteus* and the freshwater dwelling *V. niloticus* apparently lack these structures.

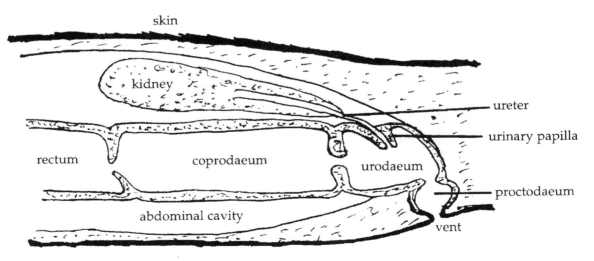

Figure 3.16. Sagittal section through the varanid cloacal region.

Each salt gland consists of numerous secretory tubules which merge into a central duct which enters the anterior chamber of the nasal capsule. The highly concentrated brine drains to the outside through the nares, or can be expelled by a sneezing-like action. The cells of the secretory tubules contain many mitochondria indicating an active transport process which removes sodium from the blood.

For an animal to achieve homeostasis (be in water balance) water intake must equal water loss. Present research indicates a great deal of variability in the water intake requirements of varanids. Like most carnivores, varanids can obtain as much as 85 percent of the needed fluid intake from the food they eat. Species such as *V. bengalensis* and *V. salvator*, which have been shown to have high water turnover rates, probably need to drink free water almost daily. Water requirements of temperate-zone varanids from Australia (*V. varius, V. rosenbergi*), which are at least partly arid-adapted, are low compared to most lizards. Their summer requirements are only half those of *V. bengalensis* and *V. salvator*. During winter, when they seldom feed and are largely inactive, they go into negative water balance, losing more water than they take in. It has been shown that the more tropical, arid adapted Australian species (*V. acanthurus*) must have access to drinking water in captivity to prevent buildup of uric acid crystals in the muscles (Wicker, 1994). However, it is not known how this finding relates to the species' life under natural conditions.

MONITOR ENERGETICS

Energy balance implies a relationship between energy intake and energy expenditure. If an energy balance is to be maintained, the energy acquired from food must equal the energy expended by the animal. If energy intake exceeds the animal's immediate needs it is stored, usually as fat. If the energy taken in is less than the animal needs, stored fat or even protein is utilized, and the animal loses weight.

On the cellular level, energy for bodily activity comes from metabolism of glucose or its equivalents. Glucose is derived from carbohydrates; equivalents for metabolism are derived from proteins and lipids. The breakdown of carbohydrates yields 4.18 Calories/gram, of lipids 9.46 C/g. Protein breakdown is more complex because nitrogenous waste siphons off 17-25 percent of the 4.32 C/g yield from protein metabolism. Since nitrogenous wastes of varanids are mostly urates and uric acid, the 25 percent loss is nearer the correct amount.

At present little is known about energy balance in varanids, except for a few species (cf Table 3.3). All species, except *V. olivaceus*, appear to be exclusively carnivorous. Thus carbohydrates, which are derived from plants, do not play a role in their food intake. Prey items contain relatively high levels of protein and lesser amounts of fat. Most prey items consumed by varanids contain 25–35 percent protein and fat. Generally 80–90 percent of the total energy present in this food can be digested and metabolized. Fur, feathers, and invertebrate exoskeletons are mostly indigestible and pass through the gut into the feces. A 20 g mouse contains about 37 Calories of usable energy.

Daily energy expenditures for an animal vary widely depending on the degree of activity. One method of determining energy expenditures is to measure it in the laboratory under restrained conditions, giving what is termed standard metabolic rate (SMR). Recently, however, many investigators have taken to the field with new techniques to measure active metabolic rate (FMR) under different natural seasonal activity regimes. Table 3.3 reflects the findings of several investigations of both types. Metabolic rate decreases as body weight increases.

We can assume from these findings that a savannah monitor weighing one kg (a full grown adult) with little activity (e.g., in captivity) would require about 22 Cal. per day, the equivalent of 0.6 mouse per day (about four 20g mice a week). In reality, it could be half this amount because of the even lower metabolic rates while asleep.

Field metabolic rates taken during the active season are, not surprisingly, much higher than SMRs, reflecting the higher energy requirements of an active animal. FMRs for varanids are about double those of iguanid lizards (Nagy, 1982), which also reflects the activity supported by the respiratory and cardiovascular specializations discussed previously.

A lizard's energy budget, such as that presented above, assumes that the lizard is neither gaining nor losing weight. If an animal is still growing, laying down fat reserves for the lean season, or producing eggs, it will have a higher

TABLE 3.3 Comparative Metabolic Rates of Some Varanid Lizards in Calories/Gram/Day (Data adapted from multiple sources)

V. acanthurus	.0049	laboratory
	.0149	field, active season
V. albigularis.	.0111	laboratory
V. bengalensis.	.0432	field, summer
V. caudolineatus	.0181	laboratory
V. eremius	.0312	laboratory
V. exanthematicus	.0219	laboratory
V. flavirufus	.0012	laboratory
	.0481	field, active season
V. giganteus	.0150	laboratory
	.0370	field, summer
V. gilleni	.0225	laboratory
V. gouldii	.0138	laboratory
V. komodoensis	.0194	field, dry season, adult
V. rosenbergi	.0285	field, summer
	.0054	field, winter
V. salvator	.0167	laboratory
	.0428	field, summer
V. varius	.0058	laboratory

energy expenditure and food requirement. It has been estimated that a female monitor needs an extra 10–15 percent average energy intake to produce a clutch of eggs.

Varanids that obtain more energy than they need for body maintenance store the reserves as fat in a pair of fat bodies located bilaterally in the posterior of the abdominal cavity (Fig. 3.11). These can weigh up to at least 10 percent of total body weight. Some species (e.g., *V. rosenbergi, V. flavirufus, V. acanthurus, V. brevicauda*) are also able to store fat in the tail, increasing its diameter by as much as two-thirds. In other species (e.g., *V. bengalensis, V. olivaceus*) caudal fat storage is confined to the interfascial spaces around the muscles at the base of the tail. These reserves can then be utilized during seasons when food availability is low. *Varanus albigularis* loses 4% of its body weight per month during the dry season of May to December.

A comparison of the energy requirements of medium-sized varanids to those of carnivorous mammals and birds of similar body weight indicates a ratio of 8 percent of the mammalian requirement and 4 percent of birds'. Varanids achieve considerable savings in energy expenditure by allowing the body temperature to drop during the night. These ratios would be lower yet for small forms because of the energy spent on endothermy in small mammals and birds. Ectotherms like varanids are thus ideally suited to exploit habitats where food is scarce, like deserts. Or conversely, they can maintain population densities much higher than could be reached by similar sized bird or mammal carnivores.

REPRODUCTIVE SPECIALIZATIONS

Male (Fig. 3.17)

Male varanids generally grow faster and reach larger size than females. Two lateral bulges at the base of the tail indicate the presence of the male copulatory organs, the hemipenes. Each hemipenis is everted through the cloacal opening when the lizard prepares for copulation, and sometimes when he is threatened and discharging the cloacal contents. Hemipenes are very elaborate structures whose morphology appears to be distinct for each species. Wolfgang Böhme (1988, 1991b) has used this distinct morphology as an aid in working out the taxonomy of varanids (Fig. 3.18). The hemipenes of varanids have two lobes, and each lobe has a terminal cartilage or bone called a hemibaculum. (Card and Kluge, 1995) The semen is transferred to the female along the spermatic groove which runs between the two lobes.

The testes are located in the abdominal cavity near the kidneys. They are relatively small except during the breeding season, when spermatogenesis is occurring. During the breeding season sperm are stored in the epididymis, a structure on the medial side of each testis. Each epididymis connects to a ductus deferens which empties into the posterior of the urodaeum close to the cloacal opening at the base of the hemipenis which has the sulcus spermaticus for seminal flow.

THE NATURAL HISTORY OF MONITOR LIZARDS

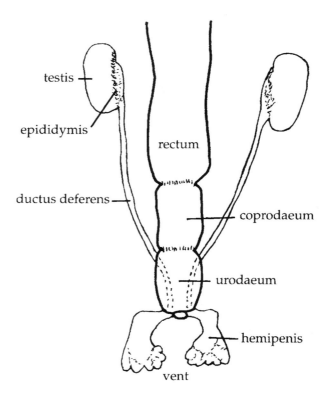

Figure 3.17. Reproductive structures of a male varanid.

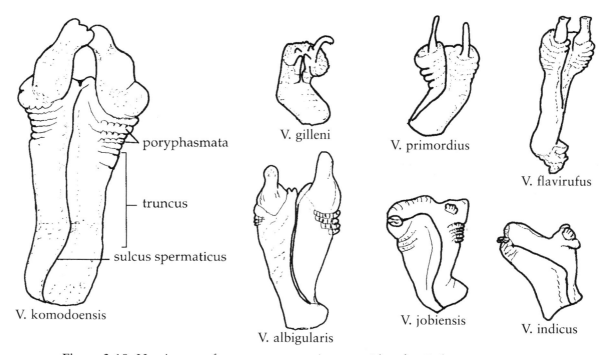

Figure 3.18. Hemipenes of some representative varanids, after Böhme (1988, 1991).

Erection and eversion of the hemipenis is accomplished by combined muscular action and blood engorgement. In copulation only one hemipenis is used at a time, the choice of right or left being determined by the side of the female on which the male is positioned.

Female (Fig. 3.19)

The ovaries are also located in the posterior of the abdominal cavity anterior to the kidneys. They are suspended in peritoneal folds. The left ovary is often larger than the right. The eggs develop from primary germ cells in the ovaries. Each egg cell is surrounded by one layer of small granulosa cells. This entire structure is called a follicle. Both ovaries normally contain the same number

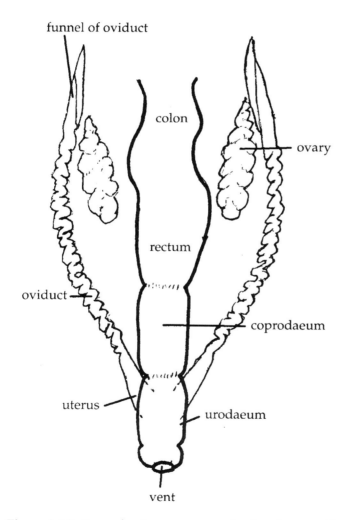

Figure 3.19. Reproductive structures of a female varanid.

of developing follicles. Follicles containing the eggs of that season's clutch begin to mature about 90 days before egg laying. During days 30-60 of development deposition of large quantities of yolk occurs (vitellogenesis). The follicles rupture when the eggs are mature. Only about half of the maturing follicles are actually ovulated; the rest are absorbed. At ovulation, they pass from the ovary into the adjacent oviduct which has moved up right next to the ovary or even partially surrounding it. The female is now ready to mate.

The oviducts have thin, almost transparent walls containing specialized glands which secrete membranes around the egg cell and yolk. Fertilization is internal with the sperm travelling through the oviducts to reach the eggs. The terminal region of most lizard oviducts is termed the uterus, although it does not function as a mammalian uterus does. Varanids have paired uteri like most lizards. After fertilization the eggs move into the uteri which contain the specialized gland that secretes the parchment-like shell over the other egg membranes. The uteri open into the cloaca. Egg development is then complete. Four to six weeks after mating, the eggs are propelled by muscular contractions from the uterus and through the cloacal opening.

In some species of varanids females have paired cloacal pouches which resemble hemipenes, although smaller, and can be everted. This can make it difficult for humans to determine lizard sex.

Chapter 4
ECOLOGY AND BEHAVIOR

DIET

All varanid lizards are entirely carnivorous, except *V. olivaceus* which is partially frugivorous. Most lizards tend toward dietary specializations, especially toward certain types of insects. Monitors, however, are extreme generalists both within and between species. The scope of the diet generally reflects prey abundance and diversity. Since the foraging technique of most varanids concentrates on finding prey rather than subduing it, a wide range of prey items is usually taken. A cockroach running from under some leaves is just as likely as not to be taken by a fairly large monitor.

Diet is thus highly variable, reflecting what is available in the habitat. It varies with seasonal prey availability. *V. olivaceus*, for example, eats mainly fruit in season for five months, and then shifts mostly to snails. Subba Rao and Kameswara Rao (1982) have tracked the dietary shift each month for *V. bengalensis* in Andra Pradesh, India, showing the species to have a definite preference for newborn rodents, even though these are not present every month and diet shifts to fish and eggs. On the other hand, Auffenberg (1994) shows that over its entire range this species feeds mostly on insects, especially beetles which have the most calories per item eaten. Diet shifts ontogenetically over an animal's life span. Hatchling *V. komodoensis*, for example, are mostly insectivores, older juveniles feed on rodents, and adults graduate to deer and wild pigs.

Diet often varies greatly in different parts of the range of widely distributed species. *Varanus flavirufus*, for example, feeds primarily on mammals and lizards in the Northern Territory (Australia). In coastal Western Australia the diet is dominated by mole crickets and spiders; in the desert interior of Western Australia reptile eggs, lizards, and beetles are the chief foods of adults (Pianka, 1994b).

Dietary studies of varanids remain largely incomplete. The usual analysis of stomach contents of museum specimens or contents flushed from live field specimens reflect the biases of season and ages of the lizards collected. However, with the exception of the four or five intensely studied species, this is the only information available. A useful method of presenting this data is the importance index, in which the volume percent of food items is calculated, preferably for a large number of lizards. This gives some idea of the dietary range of the species. The dietary importance indexes for the species that have been analyzed are presented in Table 4.1.

One important element in the diet of many, if not most, varanid species, that is not readily identifiable by stomach content analysis, is carrion. Large species are capable of tearing through the skin of large carcasses and gorging on the contents. Smaller species also are quite adept at exploiting carrion when available in proper size and condition. Road-killed small animals are

TABLE 4.1
Importance Indexes of Various Prey Types (by volume percent) of Monitor Species

Prey Type	V. acanthurus	V. bengalensis	V. brevicauda	V. giganteus	V. eremius	V. flavirufus	V. gilleni	V. exanthematicus	V. caudolineatus	V. glauerti	V. glebopalma	V. gouldii	V. g. griseus	V. indicus	V. komodoensis
Lizards	31	10	12	61	73	74	41	41	39	13	66	27	99	26	1
Snakes	—	—	—	—	—	—	—	1	—	—	—	12	—	—	—
Frogs	—	20	—	—	—	—	—	—	—	—	4	4	—	2	—
Fish	—	—	—	—	—	—	—	—	—	—	—	4	—	13	—
Mammals	—	1	—	35	—	1	—	—	—	—	—	44	—	12	98
Birds	—	—	—	—	—	—	—	—	—	—	—	—	—	—	—
Insects															
Orthopterans	43	10	16	3	17	21	28	34	14	43	30	3	1	22	1
Beetles	—	30	4	—	—	—	6	5	—	—	—	—	—	1	—
Lepidoptrans	25	—	6	—	1	2	—	9	12	5	—	—	—	1	—
other insects	—	12	13	—	—	—	—	18	—	—	—	—	—	1	—
Spiders	1	5	8	—	1	10	—	—	12	39	—	—	—	3	—
Millipedes	—	—	—	—	—	—	—	5	—	—	—	—	—	—	—
Centipedes	—	1	20	1	—	2	—	—	5	—	—	—	—	1	—
Scorpions	—	—	—	—	—	—	—	—	—	—	—	—	—	—	—
Crabs	—	4	—	—	—	—	—	—	—	—	—	2	—	1	—
Isopods	—	—	—	—	—	—	—	—	—	—	—	—	—	—	—
Mollusks	—	7	—	—	—	—	—	5	—	—	—	—	—	1	—
Fruit	—	—	—	—	—	—	—	—	—	—	—	—	—	—	—
Eggs	—	—	21	—	—	—	—	—	—	—	—	—	—	—	—

TABLE 4.1 (continued)
Importance Indexes of Various Prey Types (by volume percent) of Monitor Species

Prey Type	V. mertensi	V. mitchelli	V. niloticus	V. prasinus	V. olivaceus	V. rosenbergi	V. rudicollis	V. salvadorii	V. storri	V. tristis	V. varius	V. salvator	V. g. koniecznyi	V. flavescens
Lizards	—	—	—	—	—	24	—	44	1	74	2	8	9	—
Snakes	3	—	—	—	—	—	—	—	—	—	—	—	—	—
Frogs	2	4	5	—	—	1	54	10	—	15	—	11	1	49
Fish	2	18	—	—	—	—	—	—	—	—	—	—	—	—
Mammals	1	20	—	22	—	27	—	—	—	—	95	—	1	6
Birds	1	—	—	—	—	7	—	—	—	—	3	—	—	—
Insects														
Orthopterans	1	17	1	70	—	—	27	43	97	2	—	58	—	—
Beetles	1	1	—	—	—	—	—	—	—	—	—	—	77	—
Lepidoptrans	—	1	—	—	—	—	—	—	1	—	—	—	—	—
other insects	20	—	—	8	—	30	5	—	—	4	—	4	7	6
Spiders	—	5	—	—	—	—	7	1	—	5	—	—	—	—
Millipedes	—	—	—	—	—	—	—	—	—	—	—	6	—	—
Centipedes	—	21	—	—	—	—	—	1	—	—	1	—	1	—
Scorpions	—	—	—	—	—	—	1	—	—	—	—	3	—	—
Crabs	40	—	—	—	2	—	—	1	—	—	—	8	—	—
Isopods	—	—	—	—	—	—	4	1	—	—	—	—	—	—
Mollusks	—	—	94	—	32	—	1	—	—	—	—	2	—	—
Fruit	—	—	—	—	66	—	—	—	—	—	—	—	—	—
Eggs	29	13	—	—	—	—	—	—	—	—	—	—	4	15

readily consumed by several Australian varanids. All species tested in captivity will consume dead prey of proper size. Some monitors will seemingly go to great lengths to get at meat. *Varanus salvator* has been observed to snatch cooking meat from right over a burning camp fire.

Another significant, seasonally available, part of the diet of many monitor species is bird and reptile eggs. Since they are usually broken first and the soft parts consumed, they are retrieved from stomach contents only from live animals shortly after eating.

Cannibalism is probably a distinct possibility in most species in which lizards are part of the diet. Feeding on live members of the same species has been definitely observed in *V. komodoensis, V. giganteus, V, flavirufus, V. rosenbergi, V. storri,* and *V. griseus.*

Additional data on diet can be found in the species accounts in chapter 7.

FORAGING BEHAVIOR

Many predators attack their prey from ambush, while others usually hunt while on the move. These two strategies of foraging have entered classical ecology as the ambush ("sit-and-wait") versus the active ("wide-ranging") strategies. Lizards are often classified in one or the other, although the dichotomy is somewhat artificial. In general discussions, varanids are usually classified as active foragers, and certainly many species are. Most varanids, however, use a broad range of hunting strategies, alternately ambushing prey, searching for it in trees, burrows, and rock crevices, or digging it out of the ground. Opportunistic feeders that they are, most will also play scavenger.

Species that do forage widely as their major hunting strategy include *V. salvator, V. gouldii, V. eremius* (1.8 km/day), *V. flavirufus* (2 km/day), *V. tristis* (1 km/day), and *V. varius* (3 km/day). Large species such as *V. komodoensis* and *V. giganteus* spend at least half their day as ambush predators along game trails. Many other species alternate strategies at different times and in different habitats. *V. glebopalma,* for example, spends some hours each day exploring rock crevices for lizards, but sits in ambush for geckos at night. The highly insectivorous *V. bengalensis* uses the active technique, but only has to travel about 65 m/day to obtain sufficient food.

Most species that forage on the ground seem to have learned the best locations in their home range for finding food and return to them in regular rounds of foraging.

Monitors seem to have prodigious stomach capacities. For example, a 1.2 kg *V. varius* was found to have consumed a 500 g rabbit representing 42 percent of the monitor's own weight. Large *V. komodoensis* are thought to be able to consume 80 percent of their own weight in one meal (Auffenberg, 1981a).

Cott (1961) observed what he thought was cooperative hunting in *V. niloticus.* Observing monitors in the vicinity of crocodile nests (*Crocodylus niloticus*), one monitor appeared to distract the mother crocodile away from her nest while a second quickly dug into the nest, to be joined subsequently by the first monitor. Since this species is definitely known to be a solitary hunter, more than likely Cott was witnessing the intelligent opportunism for which varanids are noted. While the crocodile was chasing away one possible nest raider, a second seized the opportunity to dig up the eggs subsequently to be joined by other monitors in the vicinity.

There have been few attempts to study food consumption in wild lizards. Auffenberg (1994) has calculated that the average-sized male and female *V. bengalensis* require 4.6 and 5.3 g/day respectively of food to maintain weight. The same author (1984) found that satiation levels for this species were 161 g for males and 100 g for females. This indicates satiety is not reached until consumption is nearly 9 percent of body weight in males and nearly 6 percent in females.

COURTSHIP AND MATING

Courtship in most lizards is often stereotyped with display patterns distinctive at least to the genus. Varanids lack any distinctive visual displays. Being solitary animals for most of the year, monitor courtship seems primarily aimed at conveying information that the approach of the male to the female is nonaggressive.

We have complete descriptions on mating and courtship for only seven species. Pertinent descriptions follow.

In *V. komodoensis* (Auffenberg, 1981a) courtship seems to occur almost throughout the year, although actual copulation only occurs within a three month period. Frequent courtship may be related to pair bond formation. Courtship and mating often take place in the presence of small groups, such as at carrion feeding sites, among individuals with overlapping home ranges. Courtship in *V. komodoensis* begins with the approach of the female from the rear by the male, followed by the male licking the face and temporal region of the female and pressing his snout into the joint of the female's hindleg and body. The male often scratches the female on the sides and back. If the female remains stationary, the male will mount the female, crawling completely onto her back. This behavior is repeated often when these two lizards meet.

During the breeding season, courtship behavior can lead to actual copulation. The female signals her receptivity by raising the base of her tail. The male follows with a rapid tail twist, placing his vent in contact with hers. A short period of pelvic thrusting by the male follows and one hemipenis is engaged. The pair then remain still in copulation for approximately 12 min (Fig. 4.1).

In *V. rosenbergi* (King and Green, 1993), the male digs a burrow close to that of a female that he finds in breeding condition. The pair spend several days getting acquainted and courting. Courtship begins as the male enters the burrow of the female and leads her out. As in the preceding species, the male nuzzles the female and licks her head and neck. He also presses his snout around her cloacal region before mounting her. In this species the male uses his tail to raise the female's tail off the ground. The female must also raise her

THE NATURAL HISTORY OF MONITOR LIZARDS

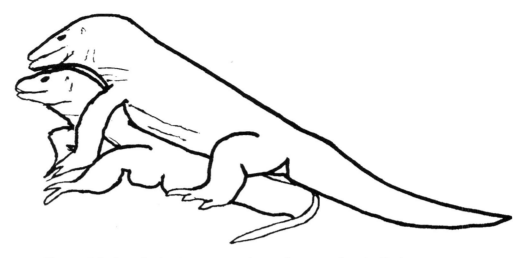

Figure 4.1. Copulation in *Varanus komodoensis*, after Auffenberg (1981a).

leg to allow their cloacas to touch. The male then inserts the adjacent hemipenis with a few pelvic thrusts. Copulation lasts about 10 minutes (Fig. 4.2), whereupon the female breaks free and reenters her burrow.

The male remains near the female's burrow, and after about a quarter of an hour reenters the female's burrow leading her out again for another copulation. This activity goes on for several days. The male uses the right and left hemipenes alternately, which seems to be a general rule for varanids (cf. Branch, 1991, for *V. albigularis*). This may be related to sperm maturation in each individual testis, as it seems to be in snakes.

Sometimes the *V. rosenbergi* pair share the same burrow during this breeding period. The period lasts several days and then the intensity of copulation declines to just a few hours in the morning. The male sometimes stays in the vicinity of the female for several more weeks, at least until after egg-laying (Green et al., 1994).

In *V. olivaceus* (Auffenberg, 1988), the male initiates the approach 93 percent of the time. Courtship in this species is very brief, about two minutes, before the male mounts the female, but does include licking and nuzzling by the male. Neck-biting of the female by the male, a rare behavior in the two previous species, is common in *V. olivaceus*. It is interpreted as an attempt to keep the female from getting away. Again, female receptivity is signaled by her lifting her tail. Copulation usually occurs only after the male has courted the female several times. Copulation lasts 2–12 min in *V. olivaceus* (Fig. 4.3).

V. timorensis is a much smaller-sized species than the three preceding (45–55 cm total length). Courtship, however, is quite similar leading to some generalizations for varanids which will be discussed below. In this species (Moehn, 1984) as in the previous ones the male almost always approaches the female from behind. Male Timor monitors immediately straddle or embrace the female and then proceed to the licking, nuzzling, and scratching behavior.

Ecology and Behavior

Figure 4.2. Copulation in *Varanus rosenbergi*, after King and Green (1993).

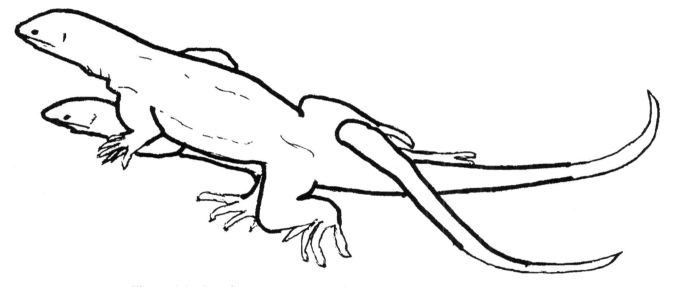

Figure 4.3. Copulation in *Varanus olivaceus*, after Auffenberg (1988).

The male attempts to keep his head aligned with the female's. Courtship can be quite lengthy in this species.

The male proceeds to use one hindleg to raise the base of the female's tail, which is accomplished only when the female is receptive. No biting has been observed. Copulation (Fig. 4.4) is achieved in typical fashion—alignment of cloacas, pelvic thrusts, period of quiet. Copulation lasts 10–15 min., and is repeated frequently.

Courtship and mating of *V. varius* in the wild has been observed (Carter, 1990). (Plate 4.1, see color section) A few differences in this species are noteworthy. This is the only species yet reported to have a visual display in courtship. The male shakes his head vigorously as he approaches the female. Females appear to mate with several males over their receptive period. Copulation between a pair is frequent, as in the other species. One pair was observed to copulate 16 times in three hours. Copulation is shorter (<4 min) than in other species so far observed, except *V. bengalensis*, in which it is reported (Auffenberg, 1988) to last only a minute and a half.

THE NATURAL HISTORY OF MONITOR LIZARDS

Figure 4.4. Copulation in *Varanus timorensis*, after Moehn (1984).

A. Brown (in Branch, 1991) describes courtship and mating in *V. albigularis*. They follow the typical pattern, although copulation may most often take place within a rocky retreat. The mean time between eight copulations witnessed in one day was 117 min.

Courtship and mating behavior, as far as it is presently known, is summarized in Table 4.2. A generalized behavioral sequence for the genus *Varanus* is as follows.

1) *Approach of the male* to the female, usually from the rear.
2) *Licking* and *nuzzling* of the female by the male, usually of the head and neck region, sometimes near the base of the tail also.
3) *Mounting* of the female by the male.
4) *Tail lift* by the female, her sign of receptiveness.
5) *Cloacal alignment* and *copulation*.
6) *Quiescent period* lasting some minutes.
7) These behaviors are repeated several times in succession.

TABLE 4.2
Summary of Courtship and Mating Behavior in *Varanus*

Behavioral acts	*V. bengalensis*	*V. komodoensis*	*V. albigularis*	*V. dumerilii*	*V. gilleni*	*V. olivaceus*	*V. indicus*	*V. rosenbergi*	*V. timorensis*	*V. varius*
Biting by male	no	no	no	yes	yes	yes	no	no	no	no
Tight embracing	yes	yes	no	?	?	no	no	no	yes	no
Hissing	no	no	no	?	?	yes	?	no	no	no
Head alignment	no	no	no	?	?	no	yes	no	yes	no
Head jerks	yes	no	yes	?	?	yes	no	no	no	yes
Licking by male	yes	yes	yes	yes	?	yes	no	yes	yes	yes
Nuzzling by male	yes	yes	yes	yes	?	yes	yes	yes	yes	no
Scratching by male	yes	yes	yes	?	?	yes	yes	yes	yes	no
Male lifts female's tail	yes	no	no	?	?	no	no	yes	yes	no
Length of copulation	1–2 min	12 min	5–16 min	?	?	2–12 min	30 sec	10–15 min	10–15 min	4 min
Female aggressive	yes	no	yes	?	?	no	no	no	no	no

The responses of the female are quite variable. She most often will do nothing — just ignoring the male's advances, or walking away. She may turn her head away in a submissive gesture. She may become aggressive—hissing, tail slapping, even biting. If she is receptive, she will lift her tail.

The role of pheromones is probably of great importance in courtship among varanids. Proctodeal (cloacal) glands seem to be important in identifying a female. Males have a great many epidermal glands along the side from the ear to behind the rear leg. These glands have very tiny pores opening in the scales. The glands themselves are simple, alveolar, and holocrine in nature and located deep in the dermis. A male *V. bengalensis* can pick up the odor of a female's trail up to three days after her passing (Auffenberg, 1994).

In species living at relatively high latitudes (20° to 30°) courtship and mating appear to be triggered solely by increased day length.

THE NATURAL HISTORY OF MONITOR LIZARDS

MALE COMBAT BEHAVIOR

It is probable that males of all varanid species engage in what is called ritual combat, although it has been observed thus far in only 14 species. This ritual combat resembles a wrestling match where the goal is to topple or floor one's opponent. Combat may involve numerous bouts before a clear winner emerges. Ritual combat usually occurs in the presence of a female.

In the larger species males begin the bout in a bipedal stance (Fig. 4.5), clutching each other with their forelegs, and using the tail for support. They twist and turn, trying to push one another to the ground. The tactic is to throw the opponent off balance. Most bouts last less than two minutes. By virtue of their heavier weight, larger males are most often the winners, but not always. After several such bouts the loser runs away, or simply flattens motionless on the ground.

Species observed exhibiting bipedal ritual combat are *V. komodoensis, V. bengalensis, V. niloticus, V. olivaceus, V. salvator, V. spenceri, V. varius, V. flavirufus, V. mertensi, V. dumerilii,* and *V. giganteus*.

In the smaller species males clutch one another with both forelegs and hindlegs (Fig. 4.6) as they roll around on the ground entwined. These bouts last much longer than those of bipedal combat — as long as 15 min. The tactic here seems to be to wear out the opponent. After a few bouts, the weaker lizard gives up and runs away.

Dominance shifts are common among similar-sized monitors and seem to be determined by regular combat bouts between neighboring males before and during the reproductive period.

Figure 4.5. Ritual combat posture between male *Varanus varius*, showing neck-arching behavior, after Twigg (1988).

Species observed engaging in ground wrestling are *V. gilleni, V. indicus, V. semiremex, V. caudolineatus,* and *V. timorensis.*

Usually biting does not occur until after dominance is determined. The winner will then sometimes bite the loser if he does not immediately retreat. Some species (*V. bengalensis* and *V. salvator*) apparently do not bite in combat at all, while others surely do (*V. dumerilii, V. gilleni, V. niloticus, V. olivaceus, V. komodoensis,* and *V. varius*).

MATING SYSTEMS

Mammals and birds have a great variety of mating systems, but reptiles have only a few. Three have been documented in varanids.

Monogamy, defined as a prolonged association between one male and one female, is the preferred system among resident adult *V. komodoensis* (Auffenberg, 1981a), and probably of *V. rosenbergi* (King and Green, 1993).

Promiscuous monogamy is defined as a male mating and guarding one primary female, but mating with other receptive females if the opportunity arises. This the system utilized by *V. bengalensis* (Auffenberg, 1994).

The promiscuous system, defined as matings with multiple partners, is exemplified by *V. varius* and *V. albigularis.*

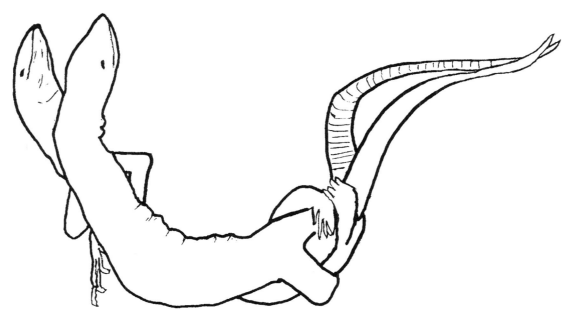

Figure 4.6. Ritual combat posture between male *Varanus gilleni*, showing clasping embrace, after Murphy and Mitchell (1974).

THE NATURAL HISTORY OF MONITOR LIZARDS

NESTING

Very little information is available on natural nesting among varanids. *Varanus komodoensis* is reported to excavate nests on hillsides. *Varanus olivaceus* is reported to use hollow trees and logs as nest sites. The most substantial observations of nesting have been reported for *V. rosenbergi* by King and Green (1993) and Green et al. (1994). The females of the latter species use active termite mounds into which they dig a 50–60 cm-long tunnel from the top (Fig. 4.7). At the end, near the center of the mound, a chamber is excavated. Excavation takes 1–2 days. *Varanus bengalensis* prefers vertical banks (Fig. 4.8) or elevated mounds for its nests, also using termitaria (Auffenberg, 1994). The female bengal monitor takes 2.5 to 3.25 hrs to dig the 25 cm deep nest tunnel. The female *V. mertensi* digs a near vertical shaft 50 cm long. After egg-laying she adds leaves and mulch to the chamber at the bottom of the shaft and firmly seals it (Ehmann, 1992).

Several other species are also reported to use termitaria for nesting: *V. flavirufus, V. niloticus, V. albigularis, V. giganteus, V. prasinus,* and *V. varius. V. salvator, V. gilleni, V. spenceri, V. gouldii* are reported to excavate nests in the ground, especially in the sandy banks of creek beds. Termite nests make perfect incubators, when available, because they protect the eggs from predators and provide nearly perfect conditions of temperature and moisture for development of reptile eggs. The termites regulate the temperature of the mound at about 30°C. Humidity in the mound is always near saturation.

Egg-laying takes a few hours, 2.5 to 4 hrs for *V. bengalensis*. *Varanus komodoensis*, in captivity at least, is known to lay over a period of several days. Afterwards, the female fills the excavation, which may also take several hours because the female rests frequently during the process. Some varanids (*V. komodoensis, V. bengalensis, V. gouldii*) at times appear to lay their eggs in communal nests. After smoothing off the surface to hide the excavation

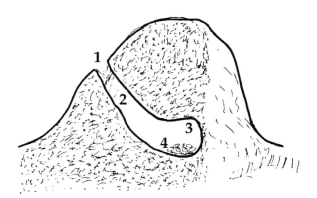

Figure 4.7. Termite mound nest of *Varanus rosenbergi*, after King and Green (1993). 1) tunnel entrance, 2) tunnel shaft, 3) nest chamber, 4) eggs.

tunnel, the female appears to mark it with scent from her cloacal region. She may pay regular visits to the nest site in the days immediately following egg-laying; it is within her home range. This may ward off diurnal predators, and perhaps prevent other monitors from attempting to use the same site or prey on the eggs.

EGGS

Considerable data is available on varanid eggs, and is summarized in Table 4.3. Eggs of all varanids have a soft, smooth, leathery or parchment-like shell. In some species the eggs adhere to one another when laid (e.g., *V. exanthematicus*, *V. salvator*), while in others they do not (e.g., *V. komodoensis*, *V. olivaceus*, *V. bengalensis*). Varanid eggs are extremely rich in lipids (13–14 percent), which provide the main energy source for the embryo during the often long incubation period (Fig. 4.9).

Clutch size is generally related to body size, with larger species having larger clutches, and larger females of a species having larger numbers of eggs. The total egg mass in a gravid *V. rosenbergi* equals 40 percent of the female's body weight. It is 21 percent in *V. bengalensis*, 23 percent in *V. salvator*, 19 percent in *V. komodoensis*, and 18 percent in *V. olivaceus*. There are, however, several exceptions to the general rule (e.g., *V. tristis*, *V. exanthematicus*)—small monitors that lay large clutches.

Fully developed eggs occupy all the available space in the body cavity of the female, putting pressure on the digestive organs. Most females cease eating for 2–4 weeks before egg-laying. The internal structure of a developing varanid egg is shown in Fig. 4.10.

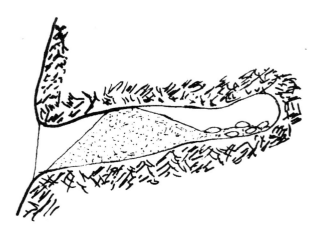

Figure 4.8. Bank soil nest of *Varanus bengalensis*, after Auffenberg (1994).

TABLE 4.3
Summary of Data on Eggs of *Varanus*

Species	Clutch size	Egg weight (grams)	Egg dimensions (mm)	Incubation period (days)
V. acanthurus	2–11	—	—	92–169
V. albigularis	9–32	—	—	116–177
V. baritji	3	—	—	—
V. bengalensis	19–30	11	40 × 24	172–254
V. brevicauda	2–8	—	—	70–84
V. caudolineatus	3–5	—	—	—
V. dumerilii	6–14	—	—	215–222
V. eremius	3–6	—	—	—
V. exanthematicus	15–40	—	—	138–181
V. flavescens	4–18	11	37 × 21	155
V. flavirufus	3–11	14	48 × 22	169–265
V. giganteus	6–1	—	—	228–235
V. gilleni	3–7	4	16 × 27	92–104
V. glebopalma	7	—	—	—
V. gouldii	6–13	34	61 × 32	—
V. griseus	10–20	—	—	300
V. jobiensis	3–5	18	25 × 54	—
V. komodoensis	2–30	160	86 × 55	208–237
V. mertensi	3–14	50	67 × 36	182–281
V. mitchelli	3–12	—	—	—
V. niloticus	20–60	—	—	129–300
V. olivaceus	4–11	44	36 × 44	180–300
V. prasinus	3–7	10	48 × 21	164–165
V. rosenbergi	10–19	26	50 × 30	180
V. rudicollis	4–16	32	58 × 32	180–184
V. salvator	6–30	50	75 × 38	85–250
V. semiremex	—	6	34 × 18	—
V. spenceri	11–35	37	56 × 31	98–130
V. storri	2–7	—	—	72–109
V. timorensis	3–11	40	28 × 17	93–186
V. tristis	5–17	—	—	93–137
V. varius	4–14	51	63 × 38	42–317

The time of egg-laying varies considerably, being tuned to environmental conditions. Temperate forms seem always to lay in late spring or early summer. Some tropical species lay their eggs in the wet season, some in the dry. Species with broad geographic ranges vary by locality. *Varanus bengalensis*, for example, produces eggs from June to September in northwestern India with the onset of the monsoons, but in December in Sri Lanka. Multiple clutches have

Ecology and Behavior

Figure 4.9. Eggs of *Varanus varius*. Photo by H. Zwartepoorte, Rotterdam Zoo.

been reported occasionally for captive females: *V. komodoensis (3); V. acanthurus (3); V. olivaceus (2); V. storri (2); V. prasinus (2); V. flavirufus.*

The incubation period is also highly variable across the genus, ranging from 60 days to over 300 days. Laboratory incubation, under constant conditions of warmth and humidity, indicates that the incubation period is partially related to body size, with longer periods for larger species. Laboratory incubation is usually shorter than that seen in the wild, probably because of low or fluctuating temperatures experienced in natural situations. Long incubation periods indicate an overwintering of the embryos inside the egg in temperate zone species, or a delay for the beginning of the rainy season in tropical forms.

Very little data is available on what proportion of the female population reproduces each year. For *V. olivaceus*, Auffenberg (1988) estimated that 90

THE NATURAL HISTORY OF MONITOR LIZARDS

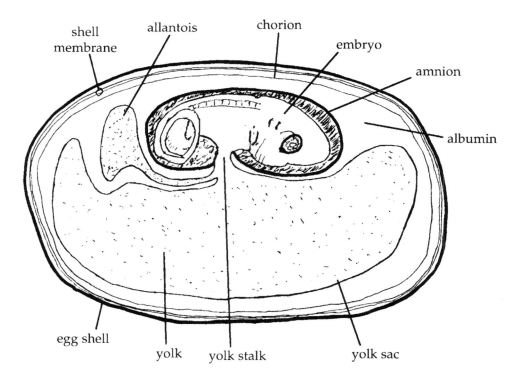

Figure 4.10. Internal structure of a developing varanid egg.

percent of all mature females lay eggs each year. For *V. flavescens* Auffenberg et al. (1989) estimate only 52 percent of mature females lay eggs each year. Auffenberg (1994) thinks that *V. bengalensis* females lay eggs every year and those in the southern part of the range may sometimes have two clutches.

During incubation the longitudinal egg diameter increases by about 50 percent and the transverse diameter by 40 percent. Mass likewise increases from absorption of moisture from the environment.

HATCHING AND HATCHLINGS

There is now a growing amount of circumstantial evidence that females of at least some varanid species return to their nests at the time of hatching to assist the emergence of the young. For those species which use termite mounds as nests, these mounds are often very hard in the center where the eggs are incubating, and no one has yet reported on the way hatchlings escape. A termite mound containing a nest of *V. rosenbergi* was plowed open during a farming operation one spring, and the hatchlings were found to have emerged

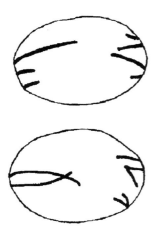

Figure 4.11. Recently hatched varanid eggs, showing slits made by egg-tooth of the baby lizard.

from the eggs but to be still encased in the hard part of the mound (King and Green, 1993). Female *V. komodoensis* may assist their young to escape the nest (Auffenberg, 1981a). Exit tunnels in termitaria nests of *V. varius* are too large to have been dug by the hatchlings, and appear to have been excavated from the outside (Carter, 1994).

Tsellarius and Men'shikov (1994) report that *V. griseus* eggs hatch between October 1–15 in the Kyzylkum desert of Uzbekistan, the hatchlings remaining in the nest burrow until the following April.

The embryo lizards grow a special structure called the egg-tooth, with which they can cut their way out of the egg shell. Slit egg shells usually indicate that the hatchlings make several attempts to open the eggs before succeeding (Fig. 4.11).

Hatchlings of different species vary greatly in size, and this data is summarized in Table 4.4. Hatchling varanids are much more colorful than adults (Plate 4.2, a–e, see color section). They usually have strongly contrasting patterns. Hatchling *V. dumerilii* have orange heads; *V. niloticus* are jet black with patterns of cream to yellow spots and dots; *V. glauerti* are iridescent blue. The reasons for the bright colors are unknown. Newly hatched varanids of most species (except *V. exanthematicus* and *V. niloticus*) are rarely seen in the wild even by professional collectors.

Hatchlings of terrestrial species are much more arboreal than adults, climbing readily. Many spend most of their youth in the trees where they are not as easily seen by predators, including cannibalistic adults. Hatchlings of many (most?) species remain together for several months, often inhabiting the same clump of trees.

TABLE 4.4
Length and Weight of Hatchling Varanids

Species	Length (mm TL)	Weight (g)
V. acanthurus	129–152	7–9
V. albigularis	200–217	23–28
V. bengalensis	180–215	14–17
V. brevicauda	80	?
V. dumerilii	179–180	16–20
V. flavescens	145	?
V. flavirufus	216–260	36–38
V. giganteus	368–381	30–50
V. gilleni	110–119	2–3
V. komodoensis	350–495	90–125
V. mertensi	252–270	23–28
V. olivaceus	345–418	25
V. prasinus	205–210	8–10
V. rosenbergi	170	18
V. rudicollis	238–260	19–21
V. salvator	280–325	21–50
V. spenceri	220	?
V. storri	117–137	2–4
V. timorensis	140–175	4–6
V. tristis	169–203	3–6
V. varius	302–372	32–40

For oviparous lizards population size and structure can be affected by (1) the number of eggs laid; (2) the number of eggs that survive; and (3) the number of hatchlings that survive to reproduce. Only the first factor has been studied for most varanids (Table 4.3).

Varanids clearly fall into the long-lived, late maturing reproductive strategy recognized by theoretical ecologists. However, the large clutch size of many species is characteristic of the reproductive strategy for short-lived animals. It undoubtedly accounts for their great success as these species have been able to resist so much commercial slaughter and environmental degradation. Hatching rate is believed to be above 80 percent in the wild in nests that escape predation. The average female *V. bengalensis* could have 400 young during her lifetime.

GROWTH

Growth rate data for young varanids in the wild is available for *V. bengalensis* (Auffenberg, 1994) and *V. albigularis* (Phillips, 1995) Auffenberg's study showed a 104 percent increase in snout-vent length in the first year, and an 8.8 percent increase in the second year of life. Phillips's study showed a 79 percent increase in SVL in the first year. Data from captive raised specimens must be used cautiously since captive monitors usually receive more food than would be available to wild young. Weight gain is especially influenced by the conditions under which the animals are kept. However, data collected by Horn and Visser (1991) and myself (unpubl.) seems to indicate that linear dimensions are not influenced by captive conditions. The meager data for linear growth are summarized in Table. 4.5.

The young of all species increase their weight by 200–400 percent during their first three months of life, and their total length by 150–200 percent. Smaller species approach their adult size in a year. Larger species require three to five years to reach adult size.

Reproductive maturity is reached in 12 months by smaller species (e.g. *V. acanthurus, V. gilleni, V. timorensis*) and at least one medium sized species, *V. flavirufus. Varanus rosenbergi, V. bengalensis, V. olivaceus* and similar fairly large species reach sexual maturity at age three. *V. komodoensis* apparently does not reach maturity until age five, and Phillips (1995) reports that *V. albigularis* in Namibia does not reach sexual maturity until age five or six. There appears to be no difference in age of sexual maturity between males and females. Young males of most species grow faster than females and attain greater overall size.

Varanids may grow continually throughout their life. This is indicated by the fact that the bone epiphyses never fuse with the adjacent diaphyses as they do in most lizards. Growth continues during adulthood albeit slowly. It can usually be said that the largest individuals are also the oldest, and the ones who were largest at maturity remain the largest. The average annual growth rate for mature *V. komodoensis* is 0.3 percent/year (Auffenberg, 1981a); that of *V. albigularis* is 7 percent/year (Phillips, 1995).

TABLE 4.5
Average Linear Growth Rate of Hatchling Varanids during First Year

Species	SV length change	Total length change
V. albigularis	96 mm	—
V. bengalensis	137 mm	—
V. f. flavirufus	140 mm	—
V. griseus caspius	62 mm	60 mm
V. rudicollis	66 mm	150 mm
V. salvator	237 mm	637 mm
V. storri	56 mm	77 mm
V. timorensis similis	36 mm	95 mm
V. tristis orientalis	41 mm	134 mm
V. varius	200 mm	291 mm

THE NATURAL HISTORY OF MONITOR LIZARDS

LONGEVITY

As a general rule, the larger the species of reptile, the longer the life span. Among most lizards the death rate is relatively high in very young age groups; middle life is characterized by a sustained period of low mortality, succeeded by a sharply rising rate in old age.

There have been no long-term field studies of marked individual varanids of any species to determine how well monitors fit the theoretical general rules. However, anecdotal evidence suggests they do fit the general pattern for lizards their size.

The only data on longevity that exists is that for captive individuals. Records of the longest period that a species has been maintained in captivity are shown in Table 4.6. Probably few, if any, individuals become senile in the wild.

TABLE 4.6
Longevity Records for Captive Varanids
(over five years)*

Species	Years
V. acanthurus	10
V. albigularis	17
V. bengalensis	22
V. dumerilii	9
V. exanthematicus	17
V. flavescens	10
V. flavirufus	18
V. giganteus	19
V. gilleni	5
V. griseus	10
V. indicus	17
V. komodoensis	16
V. mertensi	20
V. mitchelli	6
V. niloticus	15
V. prasinus	15
V. salvator	15
V. timorensis	13
V. tristis	8
V. varius	13

* all were wild caught individuals of unknown age at beginning of captive period

POPULATION DYNAMICS

Home Range

The home range of an animal is the total area it uses for all its activities through all seasons. For wide-ranging, active lizards like varanids, home ranges are comparatively large. A territory differs from home range in that it is an area which an individual defends against exploitation by others. Field studies of varanids have found no evidence of territorial behavior in any species. The ecological significance of the home range concept to non-territorial, wide-ranging animals like varanids is in some doubt. Methods of calculation also vary among researchers and can give very different results. However, since it is a regularly measured parameter in field studies, average home ranges for the species that have been studied are listed in Table 4.7.

As can be seen from the table, home range areas are highly variable. Generally, larger lizards have larger home ranges; males usually have larger home ranges than females. Home range size appears to be determined by food availability, rather than any social factor. Thus desert-inhabiting species often have huge home ranges (Phillips and Alberts, 1994), while tropical forest species ranges are tiny by comparison. Home range size of adults of the same sex often vary significantly, probably for the same reasons of resource availability.

TABLE 4.7
Mean Home Ranges for Adults of Some Varanid Species
(in hectares)

Species	Locale	Area
V. albigularis	Etosha, Namibia	(m)1800 (f)600
V. bengalensis	Sind, Pakistan	(m)5.3 (f)4.4
	Bangladesh	.133
V. flavescens	Bangladesh	.075
V. giganteus	Western Australia	(m)325 (f)47
V. griseus	Algerian desert	250
	Israel coast	(m)165 (f)319
V. komodoensis	Komodo Is.	420
V. olivaceus	Luzon Is.	1.48
V. rosenbergi	Kangaroo Is.	19.44
V. salvator	Malaya	20–120

A hierarchical system of dominance seems to determine which adult males will control the majority of matings. The core activity areas (see below) of these dominant males often overlap the home ranges of the most females. Varanids appear to know their home ranges well and to know the exact location of several retreats contained in each. Knowledge of this and other resources is the significance of establishing a home range.

Monitors lead a more or less solitary life and also live at relatively low densities. How do they maintain the social integration that the hierarchical system demands? It could be done by means of fecal scent alone. Defecation occurs most often shortly after basking, as the lizard begins to forage. Fecal pellets are deposited in the open, along trails. When a monitor comes upon such a deposit, it is investigated at great length. The lizard walks around the pellet, tongue-touches it, and pushes it with its snout for several minutes. It probably obtains information about the age and sex of the depositor. Another type of scent marking is cloacal dragging. This also occurs around shelters, and when crossing the fresh trails of conspecifics. The message may be "Pass here at your own risk" or "Females welcome, males keep out." However, it most probably is a way of establishing oneself in the local community, identifying a lizard as a neighbor not a stranger.

Core Activity Areas

The core activity area (CAA) is that part of an animal's home range in which it spends approximately 50 percent of its time, calculated usually on a daily basis. The CAAs generally include basking and sleeping sites (burrows), as well as spots with high food availability. The CAA also varies greatly among varanid species, being determined not only by foraging needs but also by season. The CAA has been calculated for only a few species. For adult *V. komodoensis*, this area is 10.8 hectares (Auffenberg, 1981a). For *V. rosenbergi*, a temperate species reaching 36°S latitude, the CAA varies greatly by season, being 0.18 ha in winter and 1.37 ha in summer (King and Green, 1993). Similar seasonal variation was found by Phillips and Alberts (1994) for *V. albigularis* at 18°S in southern Africa. The tropical, frugivorous *V. olivaceus*, which shows no seasonal pattern, has a daily activity area of only 0.05 ha (Auffenberg, 1988).

Population Density

An obvious feature of a population is that it contains a certain number of individuals inhabiting a certain area — its density. Although population density is of major interest to ecologists and conservationists, it is difficult to measure. First, it is difficult to count intelligent, wary, wide-ranging animals like varanids; and second, it is difficult to determine the area occupied. Population density is determined by two important, independent variables. One is the structural complexity of the environment; the second is the food availability.

The difficulties of determining population density are reflected in the few (8) studies to date for varanids. Auffenberg (1981a) found eight lizards per km² for *V. komodoensis*, and (1988) 61 per km² for *V. olivaceus*. Stanner and Mendelssohn (1987) found 4per km² for *V. griseus* on the coastal plain of Israel, similar to the 3–5/km² that Makayev (1982) found for this species in Tadjikistan and the 4/km in Uzbekistan reported by Tsellarius and Men'skikov, 1994). Western (1974) made estimates of 10–50/km² for *V. exanthematicus*, and 40–60/km² for *V. niloticus* in northern Kenya. Gaulke (1989) found 33 *V. salvator* per km² in the mangrove areas of Calauit Is., Philippines, while there are only 6.5/km² in the Sunderbans forest of Bangladesh (Luxmoore and Groombridge, 1990). Khan (1988) found 18 *V. flavescens* per km² in agricultural areas of Bangladesh. Phillips and Alberts (1994) counted only 0.36/km² for *V. albigularis* in Etosha National Park, Namibia. Auffenberg (1994) estimated the density of *V. bengalensis* in several different habitats over its broad range and found 39/km² around the marshes of Sind, southern Pakistan, 31/km² in an agricultural area of northern India, and 9/km² in a forested wildlife sanctuary in semiarid northwest India.

Even these few studies point out the dependence of population density on the structural complexity of the habitat. The *Artemisia monosperma* dominated sand dunes of southern Israel and the semidesert of Etosha have simple structural complexity and thus low population densities compared to the complex coastal swamp habitat of *V. salvator* in the Philippines. The effect of food availability may be reflected in the high density of the semiherbivorous *V. olivaceus*.

Biomass Estimates

Little information is available on the biomass of reptiles of any kind. The average biomass of *V. komodoensis* is 66 kg/km² (Auffenberg, 1981a) and 42 kg/km² for *V. bengalensis* (Auffenberg, 1994). Phillips and Alberts (1994) estimated the biomass of *V. albigularis* in Etosha National Park at 45,000 kg, and that of lions in the same park at 42,000 kg. The largely frugivorous *V. olivaceus* can reach an amazing 1900 kg/km² (Auffenberg, 1988).

Population Structure and Survivorship

In addition to density, a population has a structure determined largely by year-to-year survivorship of each age category. It seems certain the survivorship of hatchlings is low. Adults of larger species appear to have few predators, other than man, and survivorship in populations that are not hunted may then be high for adults. Only the long term study of *V. bengalensis* by Walter Auffenberg (1994) has yielded any real data on survivorship. He estimates a 50 percent survivorship to the end of the second year. He also estimates that only a 0.37 percent survivorship to sexual maturity (2.5 to 3 years) is needed to maintain a stable population in that species. He does not explain what happens to all the excess young. It is to be hoped that some of the long-term population studies begun in Australia will eventually yield data on population structure in other varanids.

Sex Ratio

Another population characteristic is sex ratio. The sex ratio of varanids in museum collections is heavily skewed toward a surplus of males. This bias in many species is probably an artifact of collection brought about by home range differences between the sexes. The more active males are more liable to be seen and collected. A balanced sex ratio (1:1), however, seems to be the norm for some closely studied populations (*V. acanthurus*, King and Rhodes, 1982; *V. olivaceus*, Auffenberg, 1988; *V. flavescens*, Auffenberg et al., 1989; *V. indicus*, Wickramanayake and Dryden, 1988; *V. bengalensis*, Auffenberg, 1994; *V. albigularis*, Phillips, 1995). Traeholt (1994), however, found a male biased sex ratio for *V. salvator* of 5:1 in one study site and 2.2:1 in another; and Auffenberg et al. (1989) found a male biased sex ratio of 2.23:1 in *V. flavescens*. It is very difficult to determine the sex of most varanids in the field, so few studies have yet gone into this matter. Field surveys taken during the breeding season will likely show a heavy male bias because of that sex's greater activity. Some data for *V. komodoensis* indicate a male-biased sex ratio may be present in this species at hatching. Data at present suggests that some varanids have equal sex ratios, others do not — with the exceptions favoring males.

Since some species of varanids have identifiable sex chromosomes (e.g., *V. varius*, *V. albigularis*, *V. acanthurus*, *V. niloticus*) (Fig. 4.12), and some do not (e.g., *V. salvator*, *V. tristis*), it has been suggested that temperature-dependent sex determination might be responsible for biased sex ratios in the latter two species. Breeding experiments so far have not confirmed this hypothesis.

ACTIVITY PATTERNS

Seasonal Patterns

There is great variation in seasonal activity patterns among varanid species, and even within species which have broad geographic ranges. Many tropical species are active all year, while those from temperate or desert regions are inactive during cold or dry months.

Varanus komodoensis lives in an area of the tropics with pronounced dry and wet seasons, but there is only a slightly diminished amount of activity during the dry season. *Varanus niloticus* is an African species with a large geographic range. In South Africa it is inactive during the cold season (April–August), but in the equatorial region (Senegal) it is active all year, although it does little foraging during the dry season (January–June). *Varanus olivaceus*, which inhabits rain forest with no pronounced dry or wet seasons, shows no seasonal activity pattern at all. *Varamis bengalensis* is active all year in Thailand and Sri Lanka, but hibernates from September to April in northern Pakistan. *Varanus rosenbergi*, a species occurring at a relatively high latitude (36°S), is active on sunny days all year round in its coastal habitat, but

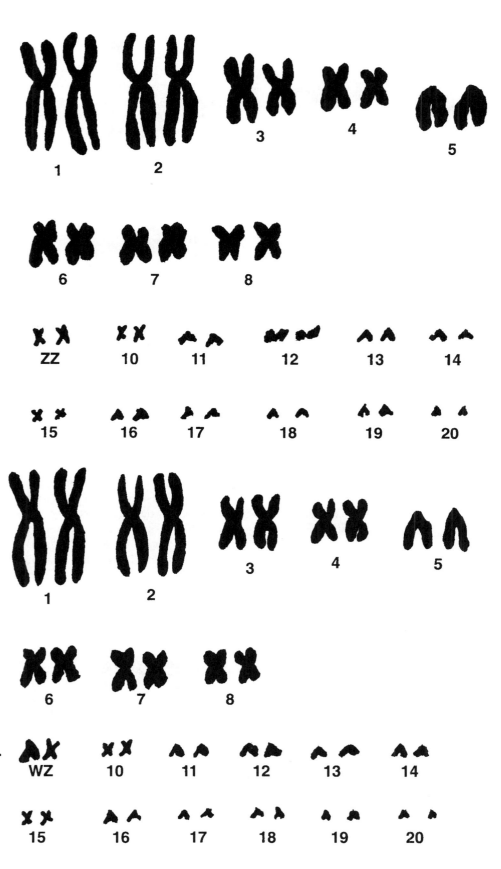

Figure 4.12. Karyotypes of male (top) and female (bottom) *Varanus varius*, after King and King (1975). Arrow indicates heteromorphic female sex chromosomes. (A karyotype is a pictorial representation of mitotic chromosomes arranged according to size.)

its activity during the winter months (May–August) is limited to the area adjacent to its burrow and little or no feeding occurs. *V. varius*, an inland Australian form, is completely inactive from May to August (winter); and *V. flavirufus* on the Great Victorian Desert is inactive for six months (March–August). Activity of *V. albigularis* during the winter dry season in Etosha, Namibia, is very limited, except for males in July–August. Surprisingly, the mating season for this species occurs during the winter. Phillips and Alberts (1994) determined the lack of activity in this species is due more to lack of food than low temperature. *Varanus griseus* in Israel has an activity season from March to November, and hibernates underground the rest of the year. In the Turkmenistan and Uzbekistan deserts *V. griseus* is active from April to October only.

Daily Patterns

The daily activity pattern of varanids can be divided into six basic phases: waking, emergence, basking, foraging, retreat, sleep. Most varanids appear to wake fairly early, but do not emerge until the sun is high enough to afford basking opportunity. Basking times vary according to ambient temperature — longer on cool mornings than warm ones.

Foraging occupies most of the day for most species. *Varanus gouldii* (Christian et al., 1994) spend up to 6 hours actively walking and foraging. *V. salvator* spends up to 13 hours a day foraging, but with some rests (Traeholt, 1994). *V. griseus* spends 11–12 hours active on the surface in May in Turkmenistan (Shammakov, 1981). It requires a study with sophisticated telemetry to determine the amount of time that monitors spend in various daily activities. Auffenberg (1988) has the most detailed account to date. He shows that *V. olivaceus* spends 5–55 percent of daylight in shelters. Christian et al. (1994) estimate that *V. gouldii* spend 64 percent of available daylight foraging, *V. mertensi* 62 percent, *V. flavirufus* 81 percent, *V. rosenbergi* 95–100 percent. However, Christian and Weavers (1994) used a sophisticated telemetry transmitter attached to the tail so that only walking and running were recorded; and this method showed *V. rosenbergi* spent an average 47.6 min/day in locomotion, *V. flavirufus* spent 97.2 min/day, and *V. gouldii* 228.5 min/day. This may reflect a lot of time eating in one spot, or more likely, considerable time spent in the ambush mode of foraging. Although most varanid species are believed to be active on a daily basis, Auffenberg (1994) found that only 10 to 12 percent of a population of *V. bengalensis* were active on any given day, with a maximum of 20 percent in the monsoon season. This type of pattern more closely resembles that of many iguanid lizards (De Lisle, 1991). It also appears that males are active more days and hours than females over a whole year. Male *V. albigularis* often walk more than 4 km in a day during the mating season (Phillips, 1995). There is evidence that foraging time is reduced during windy weather even if the temperature remains favorable. No explanation for the latter behavior is yet known.

Several species have a bimodal activity pattern during the hottest part of the year, when they are active during the early morning hours, retreat to shelter as heat builds, and then emerge again for late afternoon foraging. *Vananus bengalensis* is reported as being bimodal in arid northeastern India, but unimodal in the marshes of southern Pakistan. The meager data on varanid basic daily patterns in summer is summarized in Table 4.8.

Some authors state that varanids are all diurnal. Enough evidence is now available to say this is not true. Reliable reports indicate *V. salvator* sometimes continues to forage after sunset. *Varanus glebopalma* regularly forages for nocturnal geckos until after 2300 hrs (Sweet, pers. comm.). *Varanus tristis* is reported to forage around building lights to almost 2400 hrs, and at least one *V. bengalensis* was observed catching toads which were catching insects under a street light at 0200. *Varanus pilbarensis* forages after dark during the hot season (Ehmann, 1992). *Varanus spenceri* has been seen alert, out on a road at 2200 hrs.

Varanids usually return to their nocturnal retreat late in the afternoon. They may remain resting just outside the entrance for some time before going deeper inside to sleep. Heger and Heger (1994) report that *V. giganteus* often

TABLE 4.8
Daily Summer Activity Behavior Patterns of Some Varanids

Species	Waking	Emergence	Basking	Foraging	Retreat	Sleep
V. bengalensis	0530	0600–0800	none	unimodal (Sindh) bimodal (Punjab)	1400	?
V. eremius	?	?	?	0800–? bimodal	?	?
V. exanthematicus	?	0900	0900–0930	0930–1430 unimodal	1730	?
V. giganteus	?	0800	0800–1000	bimodal	?	?
V. glebopalma	?	0900	0900–1000	1000–2300	2330	?
V. griseus	?	0730	0730–0810	bimodal 0800–1000 1400–1800	1800	?
V. komodoensis	?	0440	0440–0615	unimodal (wet) 0615–1800 bimodal (dry)	800	1930–?
V. niloticus	?	0900	0900–1000	unimodal	1300	?
V. olivaceus	0400–0630	0700	0700–0800	unimodal 0800–1600	1740	1900–0430
V. rosenbergi	?	0850	0850–1000	unimodal 1000–1600	?	?
V. salvator	0615	0730	0730–0900	unimodal 0900–1700	1700	?
V. varius	0650	0700	0700–0800	0800–1830	1830	

use burrows for mid-day rest, but spend warm summer nights in the open in predator-free Western Australia. *Varanus bengalensis* often spend summer nights under clumps of dense vegetation instead of in their burrows.

The posture taken by varanids during sleep is different from most lizards. Even arboreal species will form a U-shape with tail draped over the head. Some species sleep with the head facing the burrow opening, some with the base of the tail nearest the opening and the head at the far end.

Varanids can construct their own burrows, or they may make use of those made by mammals or natural cracks and holes. Self-constructed burrow length is known for a few species. The burrows of *Varanus bengalensis* average 1.10 m long, with a 3.2 meter-long burrow for hibernation. The burrow of *V. rosenbergi* is about 1.25 m long, that of *V. flavirufus* 1.6 m with an additional pop-out tunnel to the surface.

THERMAL BEHAVIOR AND ECOLOGY

Monitors, like all reptiles, are primarily ectothermic. This means that they heat their bodies by basking directly in the sun or by conduction from warm surfaces. Bartholomew and Tucker (1964) demonstrated, however, that varanids can elevate their body temperature as much as 2°C above ambient temperature by endogenous heat production. This may have physiological significance on cool mornings or cool days by allowing monitors to become active more quickly. All species of varanids appear to maintain their body temperatures between 30 and 40°C from emergence until they retire to sleep. While they are foraging, most species maintain a temperature within two or three degrees of 36°C. A body temperature of 26–28°C appears to be the minimum for activity in most species.

Body size is an important factor in thermoregulation. It influences both cooling and heating rates because of the relationship between body size and surface area and heat capacity. Larger monitors both gain and lose heat more slowly than small ones. Although varanids generate more metabolic heat than most lizards, it is probably not significant in their overall heat budget.

Many lizards are very careful thermoregulators, maintaining an almost even temperature once they become active. Varanids allow their temperature to fluctuate up or down a few degrees as they move through their foraging area from cooler shady spots to hot sunny ones.

The behavior that allows most rapid heat gain is basking. Most varanids spend some time basking when they first emerge in the morning, unless the air temperature is already near their preferred body temperature, as it is on some desert summer mornings. It appears to be a common practice for a monitor to protrude only its head from its shelter upon waking, so that the sun strikes it first. It has been suggested that heating up the head first allows full mental activity levels to be achieved before the lizard ventures into the open. Horn and Visser (1988) found all the *V. giganteus* in their study area used burrows with east-facing entrances during the spring. When brain temperature reaches about 6°C above body temperature, the monitor emerges the rest of the way. Varanids will then often climb on rocks or logs to expose more of the body to the sun's rays. Burrows also aid in thermoregulation because they remain warmer at night and cooler in mid-day than the air temperature.

Activity temperature, sometimes called preferred body temperature, is determined in free-ranging lizards either by radiotelemetry or by capturing the lizard and using a thermometer. Obviously, telemetry is less disturbing of normal behavior. In the laboratory, preferred body temperature is determined experimentally in a temperature gradient. Activity temperatures of several varanids have been determined by one or more of these methods and are listed in Table 4.9.

If a monitor begins to overheat and is unable to retreat to shade or underground, it can resort to gular fluttering, sometimes incorrectly called panting. The mouth is held open and the gular area of the throat is fluttered rapidly. This increases evaporation from the mouth and reduces head temperature several degrees. Gular fluttering is triggered in laboratory conditions at temperatures above 38°C. The temperature at which a lizard loses motor coordination and is near death is called its critical thermal maximum (CTM). CTM for most species of varanids studied is about 42°C. It is 49°C for *V. griseus*.

TABLE 4.9
Activity Temperatures of Varanid Lizards

Species	Free-living (°C)			Gradient PBT (*C)
	Telemetry Mean	Range	Thermometer	
V. bengalensis			32-37	
V. caudolineatus			37.8	
V. eremius			37.5	35.9
V. exanthematicus			36.4, 34.9	36.5
V. flavirufus	35.5	27.2–38.1	37.0	37.1
V. giganteus	35.8	26.8–39.4	36.1, 36.4	
V. gilleni			37.4	37.1
V. griseus	36.8	27.0–41.6	38.5	36.4
V. komodoensis	35.1	27.6–41.3	36–40	36.3
V. mertensi			32.7	32.5
V. niloticus			32.7	34.8
V. olivaceus		27.8–38.2		
V. rosenbergi	35.6	32.1–38.0	35.1	35.2
V. salvator			27–37	35.6
V. tristis			34.8	35.4
V. varius	35.5	32.8–36.4	34.7	33.5

Skin color change can be an important thermoregulatory mechanism for many lizards—dark when absorbing heat, light when reflecting heat. Varanids, however, are not able to change skin color. There are some differences in reflectivity between species. Desert species exhibit a higher reflectivity of infrared radiation from their skin than do species from nonarid habitats.

HABITAT UTILIZATION

As we have seen, most varanids are wide-ranging and have rather large home ranges, frequently covering several types of habitat. The question arises: do they use all parts or types of habitat randomly? Ecological studies done on a few species indicate the answer is no. The findings of several of these studies serve as illustrations of habitat utilization strategies of various varanids.

In northern Australia there are places with as many as 10 species of varanids that are broadly sympatric, that is, have overlapping ranges as plotted on a map. Given their rather catholic diets and similar morphology, the question arises how do they all get along? The answer seems to be in what ecologists describe as differential microhabitat utilization. Shine (1986) indicates these northern Australian varanids show extreme habitat specialization. *V. glebopalma* and *V. acanthurus* are found only in rocky outcrops. *V. scalaris* and *V. tristis* are primarily arboreal, but occur in slightly different vegetation types. *V. gouldii* and *V. flavirufus* are widely distributed terrestrial species. *V. flavirufus* is apparently restricted to sandy soils, while *V. gouldii* inhabits areas of hard-packed ground. *V. mertensi* and *V. mitchelli* are sympatric along creeks. *V. mertensi* favors the banks of larger pools; *V. mitchelli* is usually found in the trees beside shallow, running water.

The large home range of a typical *V. komodoensis* encompasses several habitat types from quasi-cloud forest to sandy beach, but it concentrates most of its activity in the savanna and tropical deciduous forest.

V. olivaceus uses different parts of its forest habitat for different activities. It climbs to the top branches of emergent trees to bask, shelters in the vines and epiphytic growth of the canopy at night, and forages on the forest floor.

In the Indus River region of southern Pakistan, *V. bengalensis* spends over half of the activity period in the marshes and tamarisk thickets, occasionally using the *Acacia* scrubland. It totally avoids the desert scrub that is a major part of the area. *V. griseus* may replace it on the desert. In southern Sri Lanka *V. bengalensis* and *V. salvator* are sympatric (Dryden and Wickramanayake, 1991). *V. bengalensis* here is found away from the water in the tropical deciduous forest and scrub-lands. *V. salvator* occupies the riparian forest. In the Calcutta area of India *V. bengalensis* is syntopic with *V. flavescens* and *V. salvator*, but each has narrow microhabitat differences to avoid competition. The more aggressive *V. salvator* inhabits the banks of streams and riparian woodland, the least aggressive *V. flavescens* is found in isolated pools and marshes, and *V. bengalensis* inhabits the drier areas between the wetlands (Auffenberg, 1994). Where there are no other species of monitors, *V. bengalensis* uses all these types of habitat. In the Philippines, *V. salvator* uses only mangrove swamp habitat during the dry season, but also utilizes adjacent savanna during the wet season (Gaulke, 1994). Traeholt (1994) states that *V. salvator* does not range more than 50m from water in Malaya. In southern Sumatra this species utilizes all kinds of habitat, even rice fields, as long as some surface water is available (Erdelen, 1988). In Bangladesh *V. bengalensis* uses all types of habitat (farms, forests, mangroves, marshes) but may be completely sympatric with *V. salvator* in the mangroves, the only habitat utilized by the latter species (Khan, 1988).

In Estosha National Park, Namibia, *V. albigularis* is the only species of monitor, and it regularly uses all four types of habitat (grassland, grassland/bush, bush, and woodland) in about the same proportions as these habitats exist. No preference could be detected by Phillips (1995).

Elsewhere in Africa, *V. exanthematicus* and *V. niloticus* are sympatric in many regions. In Ghana (Yeboah, 1993), *V. niloticus* uses the gallery forests along rivers and streams, while *V. exanthematicus* uses the grasslands and higher woodlands.

Intraspecific competition is also usually avoided. Like domestic cats, several individual monitors may regularly use the same trails through the local habitat and share the same hunting areas, but they seem to avoid contacting one another.

AGGRESSION AND DEFENSIVE BEHAVIOR

In the wild, aggressive encounters between monitors occupying overlapping home ranges are rare. It is probable that a dominance heirarchy exists among the locals. Aggressive behavior between female neighbors is apparently nonexistant, and that among males goes no further than threat displays when they meet, except during the mating season when ritual combat may occur. If a stranger shows up in the area, a fight may ensue. Monitors can distinguish individual neighbors by their scent trails (Tsellarius and Men'shilov, 1994).

Monitors have a variety of strategies for defense. The most common is crypsis, that is, they flatten themselves against their background, become absolutely quiet, and trust their color pattern to conceal them. Most monitors will freeze in this position until an intruder is quite close. A few species have been known to feign death (e.g., *V. exanthematicus*).

When it becomes apparent that they have been spotted, monitors then usually rely on flight. The distance to which an animal will allow a potential threat to approach is called its flight distance. The flight distance for *V. bengalensis* has been calculated to be 10 m (Auffenberg, 1994). Monitors know their home range well enough to head straight for a burrow or crevice for ground and rock dwellers, the water for aquatic species, or a tree for arboreal ones. *Varanus flavirufus* has a burrow with the end near the surface, so that if it is pursued into the burrow it can pop out the far end and escape. Some of the larger terrestrial species are reported (Greer, 1989) to begin flight on four legs, but as they accelerate, raise themselves into a bipedal run on their

THE NATURAL HISTORY OF MONITOR LIZARDS

hind legs (e.g., *V. giganteus, V. gouldii, V. flavirufus*). Aquatic monitors such as *V. salvator* will swim away under the surface. Arboreal species tend to keep the trunk or a branch between themselves and the potential predator.

When startled, cornered, or attacked, the behavior proceeds to an aggressive sequence. if strongly provoked some of the larger species assume a bipedal stance (Fig. 4.13). The tail of a large varanid is a very effective weapon. Partially or completely coiling the tail horizontally usually precedes the tail slap in most species. The tail slap is a rapid lateral swing. Tail slapping has been reported for most species except the smaller arboreal ones (e.g., *V. prasinus*).

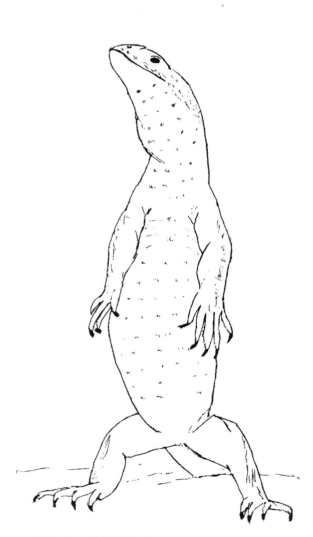

Figure 4.13. Bipedal defensive posture in *Varanus gouldii*.

In agonistic encounters with conspecifics, the aggressor performs a slow and stiff-legged walk toward the second, usually smaller, monitor. Its body is held abnormally high and laterally compressed. The vertebral column is arched and the head is held low on an outstretched, inflated neck (Daltry, 1991). (Fig. 4.14)

Gular inflation, with the hyoid apparatus expanded, seems to be used as a threat by almost all species. In species that use tail slapping, gular inflation almost always accompanies tail coiling. The common auditory act of hissing sometimes accompanies gular inflation and tail coiling, reinforcing the threat. Gaping is also used by individuals of all sizes and many species in a defensive context.

Several postural behaviors usually accompany the above actions. Lateral compression of the body permits the broadest view toward an enemy, making the monitor look bigger. A dorsal arching of the back sometimes accompanies and reinforces lateral compression. The head is then bent in a sloping position.

If physical contact with an opponent ensues, biting is used. All species of monitors will bite if physically attacked or restrained. Because of a monitor's great jaw power, its bite is particularly severe. It may be maintained with bulldog tenacity for up to 25 minutes (Auffenberg, 1988). The teeth of most varanids can inflict a deep wound which bleeds profusely, although Auffenberg states bleeding is uncommon from the peg-like teeth of *V. olivaceus*. There have been human fatalities due to hemorrhage from bites of *V. komodoensis*. If a monitor is lifted off the ground, all species make good use of their long sharp claws as weapons. There is one "last ditch" deterrent effort used in physical combat—cloacal emptying.

Figure 4.14. (A) Normal walk posture, (B) threat walk posture, after Daltry (1991).

THE NATURAL HISTORY OF MONITOR LIZARDS

PREDATION

Juvenile varanids and small species are preyed upon by a number of larger animals. The disruptive color patterns of most hatchlings indicate that predation by visually oriented carnivores may be high. Birds of prey are most likely the primary predators of small varanids, although direct evidence is lacking. Larger varanids are known to be important predators on smaller species and even young of their own species. Cannibalism has been verified in several species.

Large snakes, such as pythons and cobras, are known to feed on small varanids opportunistically. Probably most carnivorous mammals are too nocturnal or too small to be major predators of varanids. The exceptions are feral dogs and several species of mongoose.

Adult monitors of large species have rather few predators, except *Homo sapiens*, which hunts monitors mostly for their skins and sometimes for their meat. An adult *V. niloticus* was observed once doing quite well fending off a lioness with the tail slapping behavior described above.

Some monitors show only a slight fear of people, especially if no quick or aggressive movements are made. Australian species which evolved in the absence of large mammalian predators (including humans) will sometimes allow a person to come quite close to them. Such species are *V. giganteus, V. spenceri, V. rosenbergi,* and *V. gouldii*. In captivity *V. dumerilii* and *V. semiremex* are usually tame in demeanor and disinclined to bite.

Auffenberg (1994) did a study of old wounds (scars) on skins of several monitor species. He found 40 percent of adult *V. bengalensis* bear such scars. The most frequent scars are loss of toes and of tail tip. Burn scars are also fairly common. Loss of toes and burns, of course, are not due to predation but to hazards in the environment. As will be mentioned later in chapter 5, varanids seem to be particularly prone to injuring themselves. Other species in this study had a much lower scar rate: 3 percent for *V. flavescens*, 13 percent for *V. griseus*, 16 percent for *V. salvator marmoratus*, and 8 percent for *V. olivaceus*.

INTELLIGENCE

All field studies indicate that varanids have very good spatial memory—remembering places they have been. They have an inclination to follow a systematic plan when foraging. In captivity, they quickly learn where food is placed and even anticipate the actions of their keeper that mean food is coming (pers. obs.). In the wild monitors can master individualized hunting strategies that show evidence of advance planning. *Varanus komodensis* apparently keep track of the conditions of pregnant goats and horses on nearby farms, in order to be present at the birth so as to catch the newborn (Diamond, 1992). The number of published learning studies with monitor lizards is very few. Most experiments on learning in monitors have been done on feeding behavior (Krebs, 1991).

Loop (1976) showed that a monitor learned to associate food presentation with pressing a lighted switch in a significant number of trials (70%) even on the first day of the trial series. By the fourth day of trials, responses to the correct lighted switch had risen to 85 percent. This indicates that these lizards are able to solve simple problems related to foraging. They may even be more intelligent than bears when it comes to figuring out how to get at food suspended from a tree limb (Traeholt, pers. comm.).

Our human ability to learn is both remarkable and relatively unspecific. Reptile learning ability has a high degree of specificity. We can learn a lot of senseless material; monitors only learn what is important to their survival — but at that they are very good.

Chapter 5
CAPTIVE MANAGEMENT

OBTAINING HEALTHY SPECIMENS

Most monitors obtained from the wild contain a heavy burden of parasites. Tapeworms, nematodes, and protists can all take a toll on a lizard during the long trauma of capture and transport before arriving at its final destination.

Usually monitors are captured by native villagers in remote areas who fear them but need a day's pay. Several of these highly unsocial animals are then confined to a sack or screened cage where they can't escape conflict. The animal exporter only comes around to the village periodically, and the new captives are typically held without food or water in squalid conditions in the interim. The better exporters usually ship the animals out within two weeks after they arrive at the central collecting area, near an international airport. Some will even deign to provide water to the animals while they await export permits.

If the specimen you want has been brought in by one of the better importers it should have received anti-helminth injections and been rehydrated on arrival. Some dealers, however, simply operate out of the back of a truck. You should inquire from the retailer about medications the lizard has been given. In due course some monitors make it to pet stores, especially *V. exanthematicus* and juvenile *V. niloticus*. Most pet store employees are not trained in the special care needed by monitors. You may still find several crowded together inside a tank. If you can find a reptile specialist, who has hand-picked the lizards from the importer, kept them separate, and started them on a regime of medication and feeding, your chances of getting an animal worth the investment is much better.

A healthy monitor will be active and alert, with bright clear eyes. Avoid lizards that seem to be always asleep during the day. Many imported monitors arrive very thin and malnourished. Unfortunately, dealers often try to sell these animals that are little more than skin-and-bones to unsuspecting customers. It takes a lot of special care to bring a malnourished monitor back to good health. Malnourishment in monitors first shows up around the hips and base of the tail. If the pelvic bones are readily visible or the base of the tail is shrunken, it is too thin.

Avoid animals that have fecal matter caked around the vent—a sign of diarrhea, or that have soiled bellies. Healthy monitors flick their tongue at the slightest disturbance.

Many monitors are very nervous during their first few weeks in captivity. They will dash for cover if someone walks by the cage and spend a lot of time in hiding. This is not a sign of poor health; time for acclimation must be taken into consideration when obtaining a new monitor. However, keeping a monitor is not for the beginner; you should have considerable reptile-keeping experience.

Unfortunately, herpetoculturalists have not succeeded in the captive breeding and rearing of any varanids in commercial numbers. All are taken from the wild, although the numbers pale along side the thousands taken for their skins.

Almost all species do adapt to captivity after a period of acclimation lasting from a few weeks to several months. I, however, do not recommend trying to make a "pet" out of any varanid. It is true, some will tame with persistent handling and exposure to humans. However, even the large, long-term, supposedly tame captives have been known at times to turn on their owners (or their children) and inflict a bite that requires a trip to the hospital. The arboreal species in particular never tame to handling and become very stressed by handling. Remember, monitors are wild animals—they are not "puppy dogs."

HOUSING

The size and type of cage for a monitor is dictated by two considerations—adult size of the species and degree of arboreality. Most varanids show considerable capability of adapting to cage conditions that do not entirely match their natural habitat.

The all glass tanks for sale in pet stores are adequate only for those species whose adult total length does not exceed 600 mm. They can be used as *temporary* housing for juveniles of larger species. Generally, the cage for a monitor should be as long as the total length of the lizard and half as wide. The height is dictated chiefly by the degree of arboreality. However, if the door is in the top, the cage should be tall enough so that the lizard cannot make a quick exit when the door is opened. Most terrestrial species do well in a cage just high enough so they cannot push against the top. Cages for arboreal species need to be more tall than long to accommodate branches for climbing. The above dimensions are the minimum; the larger the cage you can provide, the better.

There are two basic choices for cages for larger monitors. One is the designer cage, usually made of wood with a glass front and screen top. This cage has the advantage of being very sturdy and of better controlling the microclimate (temperature and humidity). It has the disadvantage of being heavy and thus requiring a rather permanent table and place in the room. The second type is the all metal cage. Indoor galvanized steel dog kennels and steel wire rabbit cages are both inexpensive and adequate enclosures for monitors. Be sure to pick one in which the bars or wires are close enough together so that the lizard cannot get its head out (the rest of the lizard will surely follow). The major disadvantage to this type of enclosure is that it is very difficult to control the microclimate. Metal cages must be kept in a room with no drafts and one in which the temperature and humidity can be regulated.

The door of a monitor cage should be at the top. It is much safer to reach down into the cage than come face to face with the inhabitant. The door must be able to be securely fastened. Monitors have amazing strength and will push and claw against all sides of their enclosure while acclimating. Once a monitor has learned how to escape from a cage, it will do so repeatedly if the route is not blocked. Reptiles should be kept in a room closed off from the rest of the house or building, so that in the inevitable eventuality of an escape from a cage, the escapee cannot wreak havoc (literally) on the rest of your house and (psychologically speaking) perhaps on the neighborhood.

For those who have access to an outdoor enclosure, this provides the animal with a chance at the benefits of direct sunlight occasionally. Most people do not live in a climate where monitors can be kept outdoors, but they do appreciate a "vacation" outside on a sunny day. I recommend rotating monitors outdoors on the day the regular cage is cleaned. Needless to say, the outdoor enclosure must be well built and secure, but need not be fancy. It must have a place where the animal can get out of the sun if it becomes too hot.

The most practical material to cover the cage floor is newspaper. It is very practical and economical. Monitors defecate frequently and the stool is quite wet, making changes in cage substrate a frequent chore. If paper is not aesthetically pleasing, pine bark chips can be used. Another possibility is artificial turf or similar outdoor carpeting. This is quite inexpensive, provides a soft surface for the monitor to claw, and looks better, although most lizards will soon have it rolled up. It does need to be hand washed frequently, and the wear and tear of monitor's claws shred it in a few months. Also possible, if not very economical, is clean, fine to medium pea gravel.

Most monitors need some kind of artificial burrow in their cage to which they can retreat. Shelters seem to play a psychological role of reducing stress for the lizard, giving it a sense of security. The most economical type of shelter is a cardboard box with an opening cut in one end. A cardboard box can simply be discarded if it gets soiled or torn. For those who want something more attractive, any number of sculptured shelters can be purchased in pet stores, although you may have difficulty finding one large enough for a medium to large varanid.

Furnishings for monitor cages should be primarily utilitarian. Delicate objects, like plants, will be quickly destroyed. A large rock would be quite in place, and if situated under the basking light, very utilitarian. Because most monitors like to climb, at least occasionally, sturdy branches can be placed in the cage. Care should be taken to remove any snags on which the lizard might injure itself. The branches will have to be securely fastened to the cage walls or they will simply end up on the ground. Remember, monitors are stronger than lizards of other types of similar size (Figs. 5.1 and 5.2).

Since monitors are primarily diurnal animals they need a regular light/dark cycle. Provide them with 12 to 14 hours of light. Using a VitaLite® or some similar form of full spectrum fluorescent light has proved very successful in giving lizards enough UV to synthesize the vitamin D needed for proper calcium metabolism. An incandescent floodlight or spotlight (75–150 watts) is needed to provide basking warmth as well as additional light during the day. Each monitor cage must have such a basking light situated above a rock or branch at one end of the cage. Placing the light at one end, away from the shelter, provides a temperature gradient allowing the monitor to select the temperature that is best for its physiological requirements. This way one does

THE NATURAL HISTORY OF MONITOR LIZARDS

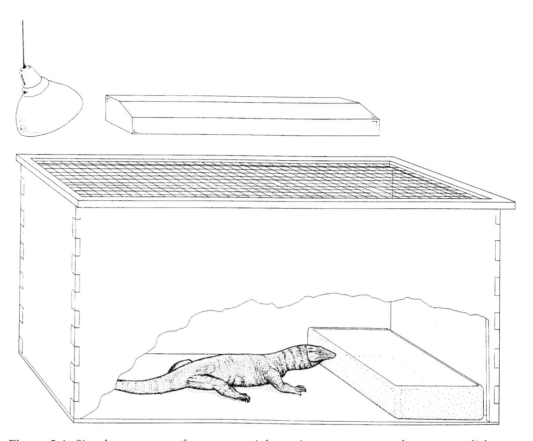

Figure 5.1. Simple cage setup for a terrestrial monitor cut-away to show water dish and lizard for scale. *Drawing by Brian Lindsay.*

not have to worry about particular cage temperatures. Care must be taken that the lizard cannot come into direct contact with any incandescent bulb, or it will get burned. Interestingly, lizards do not appear to have pain receptors in their skin as humans do to warn them of being burned.

Heat sources are best kept outside the cage. If the cage is very large or tall, a 250 watt infrared heat lamp may be used in addition to the lighting described above. Red or infrared bulbs can be left on at night where nocturnal temperatures drop below those likely to be experienced by that species in the wild (e.g., a Chicago winter). Nocturnal heat sources, if needed, for smaller cages are full of options. If the monitors are kept in a separate room, some type of room heater can be used. Keep in mind the higher cages in the room will be warmer than those near the floor. Needless to say, any room heater should be equipped with a thermostat to prevent overheating. So-called hot rocks are very popular items in reptile supply stores and catalogs. If you only have one or two cages, a carefully monitored hot rock in each can be used. Hot rocks, which each require a separate connection, are not satisfactory when multiple cages must be heated. Heat strips or small heating pads

Captive Management

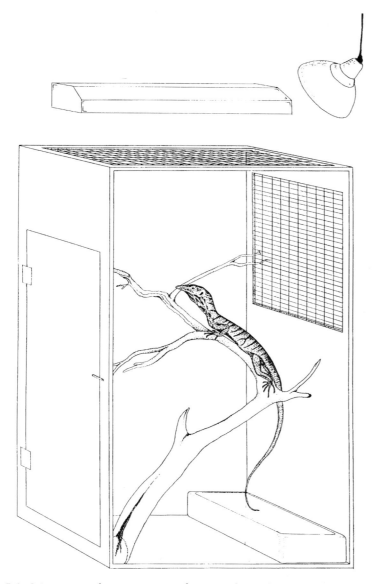

Figure 5.2. More complex cage setup for an arboreal monitor. *Drawing by Brian Lindsay.*

placed under the cages should provide sufficient warmth if the room gets moderately cold at night. Keep in mind that monitors from different climatic regimes can tolerate nocturnal temperatures down to 15°C (60°F) without any nocturnal heat source.

All cage electrical circuits should have a ground-fault interrupter of the type commonly used in bathrooms for appliances like hair dryers. Should water from an overturned dish or something similar get into the circuit, the power is turned off within milliseconds preventing electrocution.

Monitor cages need to be cleaned regularly, probably changing the substrate at least once a week. Water pans will most likely need to be cleaned and provided with fresh water daily since many monitors seem to like to soak in them for a while each day and defecate in them. It is now an issue of considerable debate as to how far one should go to disinfect a cage. There are those who argue that a mild disinfectant such as Nolvasan® or household bleach should be used to wash down the cage each time it is cleaned (Balsai, 1992). There is scientific evidence, however (Chizar, pers. comm.), that a cage which has had all its familiar odors removed places stress on the animal. It is possible that our Western cultural obsession with hygiene may be carried too far. I recommend using disinfectants only in a cage which has housed a sick lizard; soap and water can be used when a buildup of grime becomes evident. In any event, phenol-based cleansers and disinfectants, such as Lysol®, should never be used around reptiles as they are very toxic.

In handling monitors during cage maintenance it is important to immobilize both the head and the legs to prevent being bitten and scratched. The lizard should be grasped around the shoulder and neck with one hand and around the hips with the other, folding the legs backward next to the body. A pair of heavy gloves, such as welder's gloves, comes in handy for this maneuver with moderately large specimens. For specimens too long for the above maneuver, or ones where the tail must be controlled as well, safety dictates two people combine to do the handling.

Since monitors are solitary creatures by nature, it is best to keep only one lizard to a cage, except, of course, when attempting to breed them. Keeping more than one varanid together, even in relatively large enclosures, results in formation of a dominance hierarchy (Daltry, 1991). Food and basking sites are controlled by the dominant individuals. Fights break out repeatedly, frequently resulting in injuries. The subordinate lizards are under constant stress with its concomitant effects on health. It is also much easier during feeding and cage cleaning if you only have to keep an eye on one lizard at a time.

FEEDING AND NUTRITION

Because monitors are such generalists in diet, one can expect healthy monitors to feed readily in captivity once they have become acclimated. Some species that in the wild do not feed seasonally for several months (e.g., *V. rosenbergi*, *V. exanthematicus*, *V. albigularis*) or which actually hibernate (e.g., *V. griseus*, *V. giganteus*) often will stop feeding in winter in captivity as well, even though the temperatures are kept warm. This seasonality seems to be innate. Such fasts for animals that are not thin and are in otherwise good health are not a source of concern.

The kind of food to be given to particular monitors is largely driven by size. Whole-animal food is generally recognized as constituting the best diet. Because most species do feed on carrion, prekilled rodents are probably the most readily available food for those monitors that will accept them. Mice, small rats, and baby chicks form an adequate main diet. A few monitors prefer to kill their own. Although there is a slight danger of a live mouse or small rat biting a monitor, this has never happened in my collection. Monitors are quite skillful at handling live prey of proper size, very quickly learning to grasp a mouse by the side thus immobilizing the mouse's head and teeth. Horn and Visser (1989) believe, and I concur, that monitors receive some beneficial stimulation and their health is improved if they are fed living prey. A mistake that herpetoculturalists often make is attempting to feed too large an animal to a monitor; both in the wild and in captivity monitors eat a much larger dead animal than live one.

At present it is assumed that varanids have essentially the same nutrient requirements that characterize other obligate tetrapod carnivores. Thus, varanid nutrient requirements are extrapolated from the best known obligate carnivore, the domestic cat. The specific nutritional requirements associated with carnivory are as follows (from National Research Council, 1986):

    1. a limited ability to conserve nitrogen by regulation of transaminases and urea-cycle enzymes, which results in relatively high dietary protein requirements;

    2. sensitivity to a dietary deficiency of the amino acid arginine;

    3. an elevated requirement for the sulfur-bonded amino acids methionine and cystine, which are abundant in animal tissues;

    4. need for a dietary source of the amino acid taurine;

    5. the lack of the hepatic glycolytic enzyme glucokinase, which functions in carbohydrate metabolism in omnivores;

    6. the lack of the ability to utilize plant carotenes to meet vitamin A requirements;

    7. the negligible conversion of tryptophan to niacin, since niacin is abundant in animal tissues;

    8. the low rate of conversion of the essential amino acid linoleic acid to arachidonic acid, which is usually found in animal tissue.

These requirements suggest that obligate carnivores, such as almost all monitors, are best fed foods high in nitrogen, amino acids, taurine, vitamin A, and both linoleic and arachidonic acids (Allen and Oftedal, 1994). Most vertebrate prey, such as mice, will fulfill these requirements. If one were to put together an artificial diet, the suggested minimum requirements in Table 5.1 should be followed.

TABLE 5.1
Suggested minimum nutrient levels in diets of captive varanids (after Allen and Oftedal).

Nutrient	Recommended minimum level (dry basis)
Crude protein, including taurine	30–50 %
Ether extract	10–15 %
Linoleic acid	1 %
Lysine (Lys)	0.8 %
Methionine (Met) and Cystine (Cys)	0.75 %
Isoleucine (Ile)	0.5 %
Threonine (Thr)	0.7 %
Tryptophane (Trp)	0.15 %
Arginine (Arg)	1 %
Calcium (Ca)	1 %
Phosphorus (P)	0.5 %
Sodium (Na)	0.2 %
Potassium (K)	0.5 %
Magnesium (Mg)	0.04 %
Iron (Fe)	60–80 ppm
Iodine (I)	0.2–0.6 ppm
Copper (Cu)	5–8 ppm
Manganese (Mn)	5 ppm
Selenium (Se)	0.1–0.3 ppm
Zinc (Zn)	50 ppm
Thiamin	1–5 ppm
Riboflavin	2–4 ppm
Vitamin B_6	1–4 ppm
Vitamin B_{12}	20 ppb
Niacin	10–40 ppb
Folate	200–800 ppb
Biotin	70–100 ppb
Choline	1250–2400 ppm
Pantothenate	10 ppm
Vitamin A	5000–10000 IU/kg
Vitamin E	100 IU/kg
Vitamin D	500–1000 IU/kg

The best evidence at present indicates squamate reptiles are able to synthesize their own ascorbic acid (vitamin C) in the liver or kidney. Vitamin D is required in the diet only when synthesis in the skin is insufficient due to lack of UV light of the proper (285–315 nm) wavelengths. Table 5.1 assumes that the available wavelengths of light in the captive situation are insufficient for this endogenous synthesis. Even then, dietary vitamin D may be poorly utilized in the absence of UV light for basking lizards.

Most species of varanids are highly insectivorous. Some small species (e.g., *V. storri, V. timorensis*) will *not* usually eat even baby mice. Invertebrate prey are typically high in protein, but little data is available on their taurine, vitamin, or fatty acid composition. It is assumed that in the wild it is the variety of insects and other invertebrates that ensures proper nutritional requirements. Small monitors do well on a diet of crickets, but need to have it supplemented with other food. The author has had excellent results getting monitors small and large to eat turkey-based canned cat food. Another good starter or supplement is bite-sized chunks of beef heart. Most species of varanids get considerable variety in their diet in the wild. It is well to supplement the basic rodent diet with some of these other items occasionally.

Most monitors are also fond of eggs, and an occasional egg can be fed. It is advised that the egg white may be better avoided as it contains avidin which can cause vitamin B_4 deficiency if fed in excess. Eggs, however, may be a good way to get a newly acquired monitor to begin feeding during acclimation.

There are many other types of prey (food) (e.g., snails, tropical mealworms) that may be available from time to time. These also offer possibilities of diet variety. Consult the species accounts in chapter 7 for major food items eaten by wild monitors.

Monitors are not like snakes. They are active lizards that probably obtain some food almost daily in the wild. Small to moderate feedings daily or every other day are preferred to stuffing a monitor once or twice a month. Juvenile monitors especially should be fed every day to maintain growth. The quantity of food will depend on the size (species). If the lizard begins to get obese, cut back on the quantity. Adjust the diet if the lizard appears overweight. There is plenty of room for experimentation both in quantity and quality of diet. Keep records of your feeding regimen for future reference.

Monitors soon become accustomed to their feeding regimen. Care must be taken that they do not bite the hand that feeds them, or take the opportunity of an open cage door to dash for freedom. Food can be lowered through the door with a pair of forceps or tongs, if caution is used, once a monitor has reached activity temperature. I have found that placing the food in a dish in the cage early in morning before the lights are on and the lizard is fully aroused is an easy way of avoiding confrontations with your animal. Monitors quickly become accustomed to eating out of a dish. Do not introduce live prey, however, when the monitor is not ready to eat.

If you have a monitor that has become noticeably thin from a protracted fast, it may be necessary to force feed it. Because a monitor will not willingly open its mouth (except to bite), this procedure usually requires two people—one to hold the lizard, and a second to do the feeding. A soft plastic tube connected to a large diameter syringe is the best apparatus for forced feeding. Prepare a mixture of chopped beef heart, hard boiled egg, water, and a vitamin supplement in a blender, place some in the syringe, and carefully insert the free end of the plastic tubing into the monitor's mouth and far enough back so that no food is forced down the trachea. Empty the syringe slowly, allowing the lizard to swallow the mixture on its own.

WATER

Water should be available to captive monitors at all times. Even though many species from arid and semiarid habitats probably drink infrequently in the wild, captive conditions cause them to lose a lot more water through skin and respiratory evaporation. The water should be in a large, heavy container that the lizard cannot upset. Except for the few species from really arid environments (e.g., *V. griseus*) captive monitors seem to require access to water for soaking. Aquatic and semiaquatic species and those from humid tropical areas spend some time almost daily immersed in their water container when kept in the typical low humidity captive environment. These species require a large pan of water. On the other hand, never allow the cage floor to remain soggy so that the lizard has no place to get dry. Such conditions encourage skin diseases; and if ventilation is not adequate, molds will quickly develop.

BREEDING

Captive breeding of varanids is an art that is still in its infancy. Only one species, *V. acanthurus*, has been bred on a regular, although limited, basis (Wicker, 1994). The small number of breeding successes does not seem to be from want of trying. German herpetoculturalists have been trying for nearly two decades to breed various Australian species motivated at least in part by the high profits that could be realized. Records indicate that it is possible to breed monitors under captive conditions, which makes the limited success achieved so far all the more puzzling. The chief problem may be a matter of timing. Most monitors are tropical or southern hemisphere animals. Ovulation in the female seems to be triggered by a combination of photoperiod, temperature, and humidity cues which vary not only between species but also among different geographical populations of the same species. For example, *V. acanthurus* apparently breeds between July and October in various parts of Australia, while the successful breedings of this species in captivity in the temperate northern hemisphere have taken place between December and August. Female *V. bengalensis* seem to ovulate annually as soon as the number of hours of daylight exceeds the number of hours of darkness. Since females are receptive only for a few weeks, timing of mating is critical. To compound the difficulty, Wicker (1994) reports that the female can be left with the male for only one week before she begins to show signs of stress from not being able to escape the male's constant pursuit. Finally, female monitors of several, perhaps all, species are able to lay multiple clutches in a season, but probably cannot store sperm. For fertile eggs, this requires multiple matings.

Table 5.2 is a list of published records of captive reproduction. A word of caution is called for in regard to records of captive breeding published in the amateur literature. I am aware of at least two purely fictitious, fraudulent reports—one in Germany, one in the United States. Editors are requested to require proof of data contained in reports submitted by private breeders.

Analyzing the accounts in Table 5.2 allows us to formulate some guidelines that might be useful in captive breeding.

1. It seems almost trite to say it, but one must attempt to discover the sex of one's lizards. There is no certain method of sexing monitors, except surgically by an experienced veterinarian. Even the saline solution injection method requires a general anesthesia. If you have several animals it may be possible to learn it by watching their social behavior over a period of time. However, homosexual courting among monitors has been observed (Auffenberg, 1981a).

2. Monitors raised in captivity since they were hatchlings appear to breed more readily than wild caught adults.

TABLE 5.2
References of Captive Reproduction in the Genus *Varanus*

Species	Reference
V. acanthurus	Erdfelder, 1984; Wicker, 1994
V. albigularis	Visser, 1981; Van Duinen, 1983*
V. bengalensis	Klag and Kantz, 1988
V. brevicauda	Schmida, 1974*
V. dumerilii	Radford and Paine, 1989; Zimmerman, 1986
V. flavescens	Visser, 1985*
V. flavirufus	Barnett, 1979; Card, 1994
V. giganteus	Bredl and Horn, 1987
V. gilleni	Horn, 1978; Boyle and Lamoreaux, 1984
V. griseus	Igolkin (no date)*
V. komodoensis	Busono, 1974; Lange, 1989; Walsh and Rosscoe, 1994
V. mertensi	Brotzler, 1965; Eidenmuller, 1990, 1995
V. olivaceus	Card, 1994
V. prasinus	Barker, 1985
V. rudicollis	Horn and Peters, 1982*
V. salvator	Kratzer, 1973*; David, 1970
V. spenceri	Peters, 1969; 1970; 1971; 1986*
V. storri	Eidenmuller and Horn, 1985*; Stirnberg and Horn, 1981*; Bartlett, 1982
V. timorensis	Behrmann, 1981; Eidenmuller, 1986*; Sauterau and de Bitter, 1980*; Ruegg, 1974*
V. tristis	Horn and Visser, 1989
V. varius	Bredl and Schwaner, 1983; Horn and Visser, 1989

* English translation available

3. The lizards should be sexually mature and in good health.

4. Monitors should be maintained separately. Most successful breedings in the northern hemisphere have come when the natural photoperiod is lengthening and the days are getting warmer.

5. Try manipulating different humidity and temperature regimes if no courting is seen in the spring.

6. There is no evidence as to whether it is better to introduce the female into the male's enclosure, or vice versa. Observe the animals closely after introduction to make sure no fighting occurs.

7. Once actual copulations cease, the lizards should be returned to separate enclosures. Remember, the male may continue to attempt to copulate after the female is no longer receptive. She may be injured if they are not separated.

8. Egg-laying should commence within 4–6 weeks of successful mating.

9. The eggs can be incubated in the usual wet vermiculite (50% vermiculite to 50% water by weight). Probable incubation periods are listed in Table 4.2.

DISORDERS AND DISEASES

Under proper husbandry conditions monitors seldom show any signs of disease after passing through a successful acclimation period. Most disorders are either conditions that the lizard had initially when acquired or are the result of improper husbandry.

Signs of Illness

The following are signs that may indicate a health problem in a monitor.

1. Lack of alertness; eyes closed much of the time, or cloudy, or swollen.

2. Not feeding. Remember, it is normal for many monitors to feed little or not at all during winter.

3. Difficulty in breathing as evidenced by gaping or collection of mucus around mouth or nostrils.

4. Cuts, scratches, or sores should be treated at once as they can develop into abscesses.

5. Lumps or swellings under the skin.

6. Swollen toes or feet.

7. Bloody feces; diarrhea.

If signs of a disorder are detected, the question of diagnosis is next. If you are fairly sure of the origin of the problem, then you can proceed to treat it yourself if you feel you have the expertise. If the cause of the disorder is not clear or you lack the know-how to treat it, you will have to consult a veterinarian.

Finding a veterinarian with expertise in the diagnosis and treatment of reptiles may be quite difficult. In the major metropolitan areas there are usually a few veterinarians who treat reptiles. Most of their proficiency has been gained by the trial and error method, so experience is all-important. Reptile medicine is still in its infancy, and it will be necessary to make inquiries to find a veterinarian you can feel is worth the cost.

Common Disorders

Skin and foot infections are possibly the most commonly seen disorders in long-term captive monitors. Monitors kept in too humid conditions may develop discolorations of the skin. Foot abscesses can develop from the same husbandry problems. These can usually be treated with topical antibiotics as one would human skin infections. Needless to say, a drier cage environment must be provided.

Cuts and scratches are also commonly seen in monitors. Active monitors can injure themselves on items in the cage. If more than one lizard is kept in the same cage, fights resulting in injury may occur. These need to be treated with first aid to prevent infection. More serious injuries will require a veterinarian for suturing.

Less commonly, rubbed noses are seen in monitors. They are usually found in recently imported specimens. Nose rub can be part of the trauma of shipping, or it can develop from an unsatisfactory cage. If it occurs after acquiring the animal, one needs to discover the reason the animal is trying so vigorously to get out of the cage and on what it is rubbing its nose. Then corrective measures can be taken. Nose rub itself can be treated with topical antibiotics and will heal without complications. If the basic problem is not corrected, the animal can end up damaging its snout beyond repair, and death ensues.

Respiratory infections usually result from keeping monitors too cool without a proper basking site. Collection of mucus around the mouth or nostrils is a sign of respiratory disease. Gaping and gular pumping are almost certain signs of pneumonia. Monitors with respiratory infections spend most of the time with their eyes closed. Treatment should begin with raising the temperature to 32°C (90°F). Generally such temperatures can be achieved by carefully positioning the basking light and/or increasing the period it is turned on. Higher temperature increases the immune response in lizards. If no improvement is seen in a couple of days, then medical intervention is required. A regimen of injectable antibiotic is probably called for. This will usually have to be obtained from a veterinarian who will also give you the recommended dosage.

Gastrointestinal disease of bacterial origin, a major problem in captive snake collections, seems to be rare in monitors. This is not entirely surprising considering many species' predilection in the wild for putrefying meat. It is reported that feeding raw chicken pieces can lead to diarrhea, as can inadequate temperatures. If any discolored, bloody, or unusual smelling stools occur and persist for more than a few days, a veterinarian should be consulted for diagnosis as soon as possible.

Stomatitis (mouth rot) is usually seen only in recently imported specimens. As with all reptiles you are thinking of acquiring, take time to examine the mouth for any cheesy exudate, the sure sign of stomatitis. Occasionally stomatitis can be cured by cleaning and removing infected tissue coupled with a provision for a very warm basking area. Often this disease must be referred to a veterinarian, who in many cases may not be able to save the animal. Stomatitis in long term captive monitors is a sure sign of neglect of basic husbandry.

Occasionally a monitor will become infected with dry gangrene, a fungal disease which may be the result of a wound or simply unclean cage conditions. This condition usually affects the toes, and if left untreated the toes will fall off. A topical fungicide should clear it up if caught in time.

Frye (1991) states that juvenile monitors are prone to cataracts. The cause is unknown, perhaps dietary. Blindness may or may not result. There is no treatment. Puffy eyes are usually a sign of a systemic disorder rather than a problem with the eyes themselves.

Calcium deficiency is a disorder that also sometimes afflicts juvenile monitors. Its symptoms are usually swollen hind legs at first and then limb deformities. It arises from a diet deficient in the proper calcium-phosphate ratio. Crickets and baby mice are both calcium poor and should not be the sole diet of growing monitors. If a variety of food is not available, use one of the phosphorus-free supplements especially prepared for reptiles (e.g., Rep-Cal®). If the young monitors will accept turkey-based cat food, and most will, that should be fed a few times a month.

Vitamin B_1 deficiency has been found to be a problem in aquatic monitors fed almost exclusively on thawed, frozen fish. Symptoms are muscular tremors, and muscular atrophy can result. Such fish should not be a major part of monitors' diet.

Gout is the excessive precipitation of urate crystals in the joints, kidneys, or other organs. It is a relatively common and unually fatal disease of long-term captive reptiles. It is commonly thought that diets high in protein may predispose some monitors to gout. This appears to be chiefly the case where older monitors are fed heavily on protein-rich food (e.g., mice or beef heart) at cooler times of the year when metabolism is slow. Remembering that most varanids are highly insectivorous, varying the diet should avoid problems with gout.

Obesity is a common affliction of captive monitors in private collections. Monitors in small cages tend to be inactive "couch potatoes" and their owners tend to overfeed them as well. The health problems that result are similar to those of humans who overeat and don't exercise.

Parasites

In captivity, for monitors, health problems with all other disorders pale in comparison to those caused by parasites, especially internal ones. Wild monitors appear to carry a heavier parasite load than most lizards. Nothing is known of what this does to the survivorship or longevity of wild varanids.

Fortunately, external parasites are fairly easily treated—no need for a veterinarian. Ticks are the commonest external parasites of monitors. (Fig. 5.3) Several species of ticks are known only from varanid hosts (Keirans et al.,

Figure 5.3. *Varanus a albigularis* with tick infestation, Photo by K. H. Switak.

1994). Most ticks belong to the genera *Amblyomma* and *Aponomma*, which are specialized reptile parasites. I have seen a newly imported monitor with a whole colony of ticks attached to its belly scales. Unless you import directly or catch your own monitors, it is unlikely you will ever see any such infestation. Importers usually carefully scan their lizards for ticks before delivering them to the retailer. Occasionally, however, a tick may escape this scrutiny until after you have acquired the lizard. Look for ticks under the chin, between the toes, around the ear, and inside the nostrils. Apply some isopropyl alcohol (rubbing alcohol) to the tick directly. It will loosen its grip and can then be removed with a forceps.

Auffenberg (1994) has studied tick parasitism of the Bengal monitor in the wild. He never found less than 21 percent of the population with ticks, and the infestation can reach as high as 75 percent. He found an average number of 10 ticks per lizard. Larval ticks are rarely seen, perhaps living on another host. Male ticks attach mainly to the tail region; female ticks attach anteriorly, mostly around the front legs. Breeding takes place on the host lizard. Males leave their posterior attachment sites and move anteriorly where they mount the still attached females.

Although many lizards harbor red mites (chiggers), monitors do not seem to be good hosts for mites. Occasionally, a specimen can become infested with snake mites while with the exporter or in some snake-keeper's collection. Treat by placing a quarter section of 2.2 dichlorovinyl dimethyl phosphate treated pest strip on top of the cage (if it is screen) or inside the cage wrapped in nylon for 48 hours. Repeat the procedure in three weeks. Warning: prolonged or high dose exposure to pest strip can be toxic, leading to paralysis and death. Be especially careful if treating hatchlings.

Several different groups of internal parasites afflict almost all wild monitors. It is usually not necessary to aim at eliminating such parasites in newly imported monitors because the treatment is too dangerous. Limiting the burden is sufficient to guard the animal's health. Internal parasite problems almost always become apparent in the first 90 days, during which time newly acquired monitors should be kept in quarantine.

One group of internal parasites commonly afflicting monitors are the protists—single-celled organisms. Blood parasites are common protist parasites of most lizards. Monitors acquire reptile malaria (*Plasmodium*) from the transmitting insects, not mosquitos as for humans, but biting sand flies. It is not known if this has any profound deleterious effect on the lizards. Malaria has been extensively studied in lizards of the genus *Sceloporus*, and they do not seem to suffer any ill health from these blood parasites. Malaria can only be diagnosed by a complete blood exam by a veterinarian. No pathology has been demonstrated. Haemogregarines are blood parasites transmitted by mites. Again, no known pathology results.

Other protozoan parasites known to afflict monitors are *Entamoeba invadens*, an intestinal parasite which can cause ulcerative colitis. It is transmitted directly from another lizard by contaminated water. *Monocercomonas varani*, a flagellate, parasitizes the large intestine. The sporozoans *Eimeria* and *Klassia* are also gut parasites transmitted directly from fecal contamination. Importers do not usually treat for protist infections. Prophylactic treatment with Flagyl® (metronidazole) is recommended for newly imported monitors. If this drug is to be acquired from a veterinarian, he or she will probably want to examine a stool sample for a positive diagnosis. Metronidazole is a suspected carcinogen in reptiles, and must be used carefully.

The medically most important internal parasites are the worms, chiefly tapeworms, flukes and roundworms (nematodes). Tapeworms (e.g., *Protocephalus*) are exclusively digestive tract parasites as adults. The larvae of pseudophyllidean spargana tapeworms are common in the subcutaneous layer. Trematodes (flukes) can infect the stomach, gallbladder, large and small intestines. One important fluke parasite of monitors is *Reptiliotrema*.

Many types of roundworms infect monitors. Filarial worms are found in the lungs and body cavity; they are transmitted in the larval stage by mosquitos. Guinea worms (*Dracunculus*) are acquired from aquatic foods. Adult Guinea worms are subcutaneous, larval forms are found in various tissues. *Physaloptera* is a roundworm infecting the stomach, gallbladder, large and small intestine. It is acquired from eating insects infected with the larvae. *Tanqua* is a stomach nematode with a complicated life cycle involving copepods and fish. An injectable anthelmintic drug, such as ivermectin (Levamasol®), is usually given to monitors by the importer. If you acquire a lizard which has not received such an injection, you should consult a veterinarian. Since almost all worms require two different species to complete their life cycle, they are unlikely to spread further in captivity.

Recommended dosages: ivermectin 200 mcg/kg intramuscularly; metronidazole 50mg/kg orally.

Chapter 6
CONSERVATION

ENDANGERED AND THREATENED SPECIES

International law, the Convention on International Trade in Endangered Species (CITES), supposedly protects all monitors. This instrument in reality is ineffectual. Inadequate numbers of enforcement officers, lack of equipment, official disinterest on the local level, official corruption at all levels, poor education of the local people, and the high economic value of monitor skins all contribute to lax legal enforcement in spite of a multitude of laws, international and national.

In spite of the persistent trade in large numbers of varanid skins, few species are considered threatened or endangered. Four species are listed on CITES Appendix I, the instrument of international law for protecting the most vulnerable species, banning all commercial trade in the listed species. *Varanus bengalensis* is a widespread species, and is not threatened anywhere except possibly India (where all wildlife is threatened). Its listing on Appendix I is inappropriate. In spite of its listing an estimated 304,000 skins of *V. bengalensis* enter international trade annually (Luxmoore and Groombridge, 1990).

Varanus flavescens is possibly the most threatened varanid species today. An estimated 471,000 skins of this nationally and internationally protected species enter international trade annually. In spite of these horrendous numbers, it is habitat destruction that most threatens the species. It is restricted to wetland habitat of the Indo-Gangetic plain which is rapidly being drained and converted to agriculture.

In the international meeting of CITES in November 1994, the populations of *V. bengalensis* and *V. flavescens* in Bangladesh were removed from Appendix I and placed on Appendix II with other monitors.

The third species listed on CITES Appendix I is *V. griseus*, another widespread species that is little threatened through most of its arid range. Even the skin trade, which pays little attention to CITES, has not favored this species (only 20,000 reported 1975–86). There is some concern for the subspecies *V. g. caspius* which has seen much of its habitat converted to agriculture in the last 30 years. It is still locally common, however, in Turkmenistan (Makayev, 1982).

The fourth species listed on CITES Appendix I is *V. komodoensis*. Although of very limited distribution, fortunately for the komodo monitor its scales contain osteoderms (large bony plates) making its skin unsuitable for manufacture into leather goods. In addition, *V. komodoensis* occurs mainly in a national park where it is a prize tourist attraction and does receive adequate protection.

These same four species have been placed on the U.S. Endangered Species List without consultation with the herpetological community and without the support of scientific data.

All other species of *Varanus* are listed on Appendix II of CITES, requiring special documentation permits for international trade.

The IUCN Red List of Threatened Animals (a nonlegal but highly respected conservation document) also lists four species of varanids. *V. flavescens* is listed as Threatened; *V. griseus caspius* is listed as Vulnerable; *V. olivaceus* and *V. komodoensis* are listed as Rare.

THREATS TO MONITORS

Monitors have a long history of exploitation by humans. Many African, south Asian, and Australian traditional cultures still use varanids for food. The Australian desert aborigine recipe is: break neck, throw onto coals, eat when skin begins to peel. The people of New Guinea and Solomons considered the fat bodies a gourmet delicacy. Most of these same cultures used monitor parts in their folk medicine. Varanids also figure in Australian aboriginal mythology.

Skins of varanids have been used on a limited basis in traditional cultures for millennia, mostly in the manufacture of drums. It is the late twentieth-century leather trade, however, that poses the most serious threat to monitor conservation. The trade in monitor skins is very extensive.

All evidence suggests that exploitation for the skin trade tends to depress population levels, and whenever trapping is carried out regularly for extended periods it can result in local disappearance of the species. This has occurred with *V. salvator* in Indonesia, for example, at localities within easy reach of main towns and trade centers. About 700,000 skins of this species were exported from Indonesia in 1987. Widespread disappearance of *V. salvator* has occurred in the Philippines due to sustained exploitation for the skin trade. In 1985 alone, over 80,000 skins from the island of Mindanao of the brightly colored *V. s. cumingi* were exported to Japan.

Varanus salvator collected for their skins are hunted by means of baited-noose traps of various kinds. *V. niloticus* are caught with hooks baited with fish or frogs. *V. bengalensis* and *V. flavescens* are usually caught by hand after excavating their burrow, or are hunted with the aid of dogs.

The main countries importing varanid skins are Japan, France, Italy, and the United States. Japan alone accounts for nearly 50 percent of the imports. Although a signatory to CITES, Japan has taken a reservation with regard to the monitors on Appendix I and continues unrestricted trade in *V. flavescens* and *V. bengalensis*. There is often poor agreement between the number of skins supposedly exported and those reported as being imported.

Almost all the commercial trade in skins is derived from seven species. Table 6.1 lists the estimated annual production of *Varanus* skins entering international trade.

Items manufactured from these skins are chiefly shoes, wallets, belts, handbags, and watchstraps.

According to Djasmani and Rifanie (1988), a group of three professional hunters can catch up to 40 *V. salvator* per day by setting some 200 snares. They make about US$3.50 per skin if they prepare them themselves. *Varanus bengalensis* skins fetch an average of US$3.00. Because of their coarse grain and small size, *V. flavescens* skins only bring about US$1.75.

The most successful strategy to reduce the number of monitors killed for their skins is probably to reduce the demand and thereby reduce the market value of the skins. Using the same publicity techniques that have been used successfully to reduce the demand for furs and ivory should bring results. Making lizard skin products socially unacceptable would be a wonderful long-term project for regional herpetological societies.

In theory, commercial exploitation of varanids could be conducted on a sustained yield basis. No studies have yet been conducted to determine what a sustained yield harvest would be for any species. At present, exploitation for skins is recognized as seriously depleting populations wherever it is occurring.

By comparison, the number of varanids reported in trade as live animals is minuscule. A wider variety of species is caught, but for only a few is the take in numbers likely to be significant ecologically. Interestingly, some of the species that are most important in the skin trade also form the bulk of the live animal trade, presumably because the common species are better suited for supplying a bulk pet market. Table 6.2 lists the total minimum net trade in live varanids for the last decade.

TABLE 6.1
Estimated Annual Take in Monitor Skins for International Trade

V. bengalensis	304,000
V. exanthematicus	57,000
V. flavescens	471,000
V. griseus	17,000
V. indicus	2,000
V. niloticus	406,000
V. salvator	1,530,000

TABLE 6.2
Minimum Net Trade in Live Varanids Recorded in CITES Annual Reports 1981–1990

Species	Total traded (10 years)
V. acanthurus	15
V. bengalensis	1253
V. dumerilii	2150
V. eremius	22
V. exanthematicus*	114,964
V. flavescens	122
V. flavirufus	69
V. gilleni	19
V. gouldii	40
V. griseus	29
V. indicus**	1,166
V. jobiensis	25
V. komodoensis	14
V. mertensi	3
V. mitchelli	3
V. niloticus	21,709
V. olivaceus	14
V. prasinus***	11,739
V. rudicollis	1,808
V. salvadori	53
V. salvator	39,300
V. storri	4
V. timorensis	1,325
V. tristis	61
V. varius	14

* *V. albigularis* has not been separated from *V. exanthematicus*, although most *V. albigularis* entering the trade have done so since 1990.
** Includes *V. doreanus*.
*** Includes *V. beccarii*.

Humans have sometimes used varanids in unusual ways. The Japanese are said to have introduced *V. indicus* to many parts of Micronesia to control rats. Australians are said to have introduced *V. rosenbergi* to some off-shore islands to control venomous snakes.

Tens of thousands of monitors are accidentally killed on highways, chiefly in Australia and increasingly in southern Asia. Some species are killed deliberately in large numbers because they are considered bad luck or because they eat chickens and their eggs.

However, next to the skin trade the major negative impact humans have on varanids is the massive destruction of their habitat that is occurring in Asia, Africa, and even Australia. Forest-dwelling species (e.g., *V. dumerilii* and *V. rudicollis*) are probably most at risk because huge areas are being cleared by logging. The large and rapidly increasing human populations of Africa and southern Asia are exploiting the natural resources at an increased rate, and there is continual demand for more land. It is chiefly this largely poor and often hungry human population pressure that has frustrated all legal efforts to protect wildlife in India, even thwarting the presumed safety of national parks and wildlife preserves. The conservation ethic is either a totally alien concept or is of little or no concern to poor villagers when they see opportunities to obtain more food, money, or land. Most monitors live in areas where the vast majority of the population lacks adequate education to cope with the great changes in land-use patterns, culture, and technology that afflict all of us living at the end of second millennium.

Significantly, studies have shown that conversion of land to agriculture does not necessarily negatively impact many varanid species (e.g., *V. salvator*, *V. griseus*, *V. bengalensis*) *provided* the lizards are not hunted by the local populace (Luxmoore and Groombridge, 1990; Makayev, 1982). Agriculture frequently brings more food to monitors (orthopterans [grasshoppers, locusts] and rodents) and the monitor population can climb significantly, if a hunting ethic is not present among the local humans.

The effects of pollution by industrial wastes and agricultural chemicals on monitors is unknown. Most developing countries have few controls on waste disposal, and agricultural pesticides banned in Europe and the United States a generation ago are still available and utilized in the developing countries. Since monitors are carnivores and near the top of the food web, it can only by assumed they are affected by accumulating levels of toxins derived from their prey.

Urbanization is undoubtedly having a negative impact on varanids in those areas where human population growth is most rapid.

REGIONAL PROBLEMS

Africa

The desert monitor, *V. g. griseus*, is widespread over most of North Africa. Although population densities are low, exploitation is also low. Some are taken for use in folk medicines. Habitat remains fairly intact and the species is judged to be in no danger.

The Nile monitor, *V. niloticus*, has long been sought for its skins. Today the skin trade is concentrated on populations in five countries—Cameroon, Chad, Mali, Nigeria, and Sudan. The spectacular Lake Chad population is under intense hunting pressure. Local extinctions can be expected, but the species as a whole is not considered threatened because it is so widespread.

The white-throated monitor, *V. albigularis*, is a fairly large monitor that would seem to be a candidate for the skin trade, but has apparently escaped being taken in large numbers so far. Its populations are under pressure of habitat destruction in parts of East Africa.

The savanna monitor, *V. exanthematicus*, does enter the skin trade, chiefly from Nigeria and Sudan. Heavy demand by the pet trade for all sizes of savanna monitors, together with habitat destruction, may be causing serious declines of this species in a few areas, but the species as a whole does not appear threatened.

Most Nile monitors and savanna monitors entering the pet trade are exported from Togo and Benin but probably originate from neighboring countries as well. Live white-throated monitors are exported in small numbers from Tanzania and Zambia.

Asia

The conservation status of some of the Asian varanids is of greater concern than the African species. There is a long history of exploitation of monitors for food and skins in several Asian countries, and human population pressures and habitat destruction have become extreme in some countries.

The yellow monitor, *V. flavescens*, from the Indo-Gangetic wetlands is heavily exploited for its skins and habitat destruction in India is severe. It is a candidate for Threatened status.

With over a million water monitors, *V. salvator*, being killed each year for their skins, many populations of this species have shown serious decline, especially in Indonesia and the Philippines. However, because of its large geographic range and ability to survive around the edges of urban centers, it is not considered threatened as a species. The taxonomy of this species needs serious review. Many distinct insular varieties exist, and some of these may be in greater danger than the species as a whole.

Populations of the Bengal monitor, *V. bengalensis*, although fairly heavily exploited for their skins, appear to be in good shape except in India where hunting of all monitors continues throughout the country in spite of legal protection.

Several other Asian monitor species are exploited for the live animal trade. The only species with significant numbers exploited live is the water monitor, but even these numbers are insignificant compared to those taken for the skin trade. Most Asian monitors exported live come from Indonesia, which is also the chief exporter of skins. Bangladesh and Thailand rank second and third in skin exportation. A few live monitors are also exported from Malaysia, Solomons, and Vietnam.

None of the other species of Asian monitors is considered threatened at this time, although some subspecies may be. Insufficient data has been gathered to render a judgment on their status. Some rare forms, such as Gray's monitor, *V. olivaceus*, are highly localized and need constant monitoring to foresee the intense pressure of human population growth and deforestation occurring on many islands. New Guinea especially has several endemic varanid species. Logging pressures are just beginning to be felt on this island paradise. Vigilance will be necessary to head off the destruction that has visited other parts of the southeast Asian rain forest.

Australia and Oceania

The extremely strict, some would say overly strict, laws and regulations of Australia are vigorously enforced. There is no commercial trade in skins or live animals of any of the Australian varanids. Collecting or keeping varanids is permitted only for research purposes or by approved zoos. The species and numbers on Table 14 reflect the latter. Only Australian aborigines are permitted to take monitors for food. Australian herpetologists are generally not permitted to keep monitors, even ones caught in their own back yard.

Road traffic does take a fair toll of varanids, especially the carrion-feeding species. Drawn to road-killed carcasses of other animals on the highway, monitors are themselves hit by oncoming vehicles. The numbers are sufficient along main Queensland highways to have an impact on local populations, as has occurred with many snake populations in the American West.

Habitat destruction is also affecting some local varanid populations, especially in New South Wales and eastern Queensland. Most Australian monitors, like their Asian congeners, could probably survive much habitat alteration, but dogs and especially cats seriously reduce local populations near human habitations. This is not noted at first, because the impact is mostly on hatchlings and the adult population may remain stable for many years. The imported marine toad, *Bufo marinus*, is known to eat small monitors, and may be increasing the pressure on monitor populations on the Queensland coast.

At present, the conservation status of all Australian species is considered sound (King and Green, 1993).

Varanus indicus populations in the southern Mariana Islands are reported (McCoid et al., 1994) to be in decline in developed areas where the species is often a traffic victim. It is also considered an agricultural pest because of its reputation for raiding chicken coops, and it is shot and poisoned.

CONSERVATION LAWS AND EDUCATION

Perhaps no better example exists of the futility of relying on legislation to protect wildlife than what has happened to India's monitors. The Wildlife Protection Act of 1972 and 1977 gives special protection to all four monitor species. Hunting, trade, and export of the species and products are prohibited. In the two decades since legal protection, hunting has continued unabated so that monitors are now found only in a few areas, mostly in the western part of the country (Gupta, 1994). Trade in monitor skins across the border to Bangladesh occurs on a large scale.

This perhaps extreme example illustrates the general rule that all governments tend to put more emphasis on legal considerations to the neglect of, and even sometimes to the exclusion of, ecological considerations. This has caused actual detriment to the species in some cases, e.g., *V. flavescens* in India.

The root causes of population declines in varanids are considerably more difficult to address than just passing laws. The areas of growing human population correlate rather well with the range of the genus *Varanus*. This growth places increasing demands on natural resources and ecosystem processes that are already impoverished and stressed. Settlement policies promote the movement of the growing unemployed labor forces to areas where conflict with wildlife is inevitable.

Current conservation efforts are insufficient to reduce the trend toward population declines in varanids. Traditional means, such as national parks, programs for species of special concern, zoos, and captive propagation are key elements of conservation and will remain so. Indeed, they are necessary and must be greatly strengthened, especially protection of national park lands and restructuring laws to include more of the private sector in captive propagation.

New dimensions must be added to the traditional set of approaches. Economic activity derived from wildlife resources must be linked with the welfare of rural people in surrounding regions. To many citizens, the goals of conservation and use seem diametrically opposed—mutually exclusive. This is at best misguided and at worst the result of animal rights and welfare activists advancing their causes even when they have no conservation merit (Webb, 1994). The high reproductive rate of most varanids indicates that a sustained harvest of juveniles for the pet trade could be an economic asset to the local community and in turn lead to valuing the importance of the adult breeding population. This has already begun to occur in some west African countries for *Python regius* (Greer, 1994).

Education is another key element that must be emphasized in the future. Life science courses are presently organized around a systematic survey of cells, physiology, and the taxonomic hierarchy of living things, culminating in a close inspection of human biology. Ecology is frequently relegated to a scant end unit forced into the final week or two of the school term. Even in environmental education programs where ecological interdependence is supposedly the central theme, wildlife population biology is treated superficially if at all.

Especially in developing countries, teachers must be trained to take advantage of students' natural fascination with spectacular animals like varanids. This can lead to considering the aesthetic and scientific values of these wild populations and developing arguments for their protection.

Ecological considerations should have a place in virtually every discipline, not just life science. To be truly effective in promoting a conservation ethic, ecosystem thinking cannot simply be tacked onto existing curricula; it must pervade them. Promoting stewardship and sustainable use of their nation's biological resources must become as central to the purpose of the educational system as is the development of a literate, skilled work force.

Plate 4.1. Mating pair of lace monitors, *Varanus varius*. Courtesy of J. van der Koore, Rotterdam Zoo.

Plate 4.2. Hatchling monitors, showing bright colors found in many species:
a. White-throated monitor (*V. albigularis*). Courtesy of D. Northcutt.

4.2 b. Dumeril's monitor (*V. d. dumerilii*). Courtesy of K. Tepedelan.

Plate 4.2 c. Nile monitor (*V. niloticus*). Photo by author.

Plate 4.2 d. Spencer's monitor (*V. spenceri*). Photo by K. H. Switak.

Plate 4.2 e. Timor monitor (*V. t. timorensis*). Photo by W. B. Love.

Plate 7.1 a. *Varanus acanthurus brachyurus*, Mt. Isa, Queensland, Australia. Courtesy of R. W. Van Devender.

Plate 7.1 b. Habitat of *Varanus a. brachyurus* and *Varanus giganteus*, vicinity of Alice Springs, Northern Territory, Australia. Courtesy of R. Hoser.

Plate 7.2 a. *Varanus albigularis albigularis*, South Africa. Courtesy of P. Zupich.

Plate 7.2 b. Habitat of *V. a. albigularis*, Mountain Zebra National Park, elevation 1300 m., Cape Province, South Africa. Photo by K. H. Switak.

Plate 7.3. *Varanus albigularis ionidesi*, southeastern Tanzania. Courtesy of N. Miner.

Plate 7.4. *Varanus albigularis microstictus*, Tanzania. Courtesy of D. Northcutt.

Plate 7.5. *Varanus beccarii*, Aru Islands, Indonesia. Courtesy of D. Northcutt.

Plate 7.6. *Varanus bengalensis bengalensis*, Pakistan. Courtesy of A. Toth.

Plate 7.7. *Varanus bengalensis irrawadicus* (holotype), Wanding Valley, Yunnan, China. Courtesy of Yang Datong.

Plate 7.8. *Varanus bengalensis nebulosus*, South Vietnam. Courtesy of D. Northcutt.

Plate 7.9. *Varanus bengalensis vietnamensis* (holotype), northern Vietnam. Courtesy of Yang Datong.

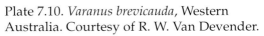

Plate 7.10. *Varanus brevicauda*, Western Australia. Courtesy of R. W. Van Devender.

Plate 7.11. *Varanus caudolineatus*, Western Australia. Courtesy of R. G. Sprackland.

Plate 7.12. *Varanus doreanus doreanus*, Irian Jaya, New Guinea. Courtesy of D. Northcutt.

Plate 7.13. *Varanus dumerilii dumerilii*, Kalimantan (Borneo), Indonesia. Courtesy of D. Northcutt.

Plate 7.14. *Varanus eremius*, Olga Mountains, Northern Territory, Australia. Courtesy of G. Visser.

Plate 7.15. *Varanus exanthematicus*, Togo. Photo by author.

Plate 7.16. *Varanus flavescens*, Sind, Pakistan. Courtesy of D. Northcutt.

Plate 7.17 a. *Varanus flavirufus flavirufus*, Shay Gap, Western Australia. Courtesy of R. Hoser.

Plate 7.17 b. Habitat of *V. flavirufus* and *V. gilleni*, Simpson Desert, South Australia. Photo by author.

Plate 7.18. *Varanus flavirufus gouldii*, Clare Valley, South Australia. Courtesy of R. W. Van Devender.

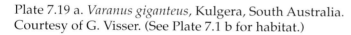

Plate 7.19 a. *Varanus giganteus*, Kulgera, South Australia. Courtesy of G. Visser. (See Plate 7.1 b for habitat.)

Plate 7.19 b. *Varanus giganteus*, Australia. Courtesy of R. W. Van Devender. (See Plate 7.1 b for habitat.)

Plate 7.20. *Varanus gilleni*, Simpson Desert, Northern Territory, Australia. Photo by K. H. Switak. (See Plate 7.17 b for habitat.)

Plate 7.21. *Varanus glauerti*, Kimberley, Western Australia. Courtesy of G. Visser.

Plate 7.22. *Varanus glebopalma*, northern Australia. Courtesy of D. Frances. (See Plate 7.30 b for habitat.)

Plate 7.23. *Varanus gouldii gouldii*, Kakadu N. P., Northern Territory, Australia. Courtesy of D. Northcutt.

Plate 7.24. *Varanus gouldii horni*, Irian Jaya, New Guinea. Courtesy of D. Northcutt.

Plate 7.25. *Varanus griseus griseus*, Egypt. Courtesy of D. Frances.

Plate 7.26. *Varanus griseus caspius*, Karakum Desert, Turkmenistan. Courtesy of T. Papenfuss.

Plate 7.27. *Varanus griseus koniecznyi*, Pakistan. Photo by author.

Plate 7.28 a. *Varanus indicus*, Guadalcanal, Solomons. Courtesy of D. Northcutt.

Plate 7.28 b. Habitat of *V. indicus* and *V. prasinus*. Sau River, Papua New Guinea, elevation 700 m. *Varanus indicus* is found near the river's edge; *Varanus prasinus* in the forest away from the river. Photo by K. H. Switak.

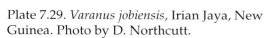

Plate 7.29. *Varanus jobiensis,* Irian Jaya, New Guinea. Photo by D. Northcutt.

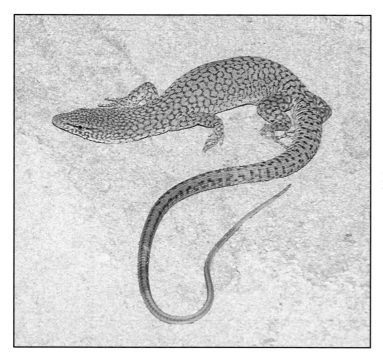

Plate 7.30 a. *Varanus kingorum*, Halls Creek, Western Australia. Courtesy of R. Hoser.

Plate 7.30 b. Habitat of *V. kingorum* and *V. glebopalma*, vicinity of Lake Argyle, Western Australia. Courtesy of R. Hoser.

Plate 7.31. *Varanus komodoensis*, Flores Island, Indonesia. Courtesy of R. W. Van Devender.

Plate 7.32. *Varanus mertensi*, Mount Isa, Queensland, Australia. Courtesy of G. Visser.

Plate 7.33 a. *Varanus mitchelli*, Kurrarurra, Western Australia. Courtesy of R. Hoser.

Plate 7.33 b. Habitat of *V. mitchelli*, Ord River, Western Australia. Courtesy of R. Hoser.

Plate 7.34 a. *Varanus niloticus*, Togo. Courtesy of R. W. Van Devender.

Plate 7.34 b. Habitat of *V. niloticus*. Crocodile River, vicinity of Kruger National Park, South Africa. Photo by K. H. Switak.

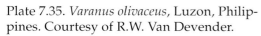

Plate 7.35. *Varanus olivaceus*, Luzon, Philippines. Courtesy of R.W. Van Devender.

Plate 7.36. *Varanus pilbarensis*, Pilbara, Western Australia. Courtesy of D. Frances.

Plate 7.37. *Varanus prasinus*, Irian Jaya, New Guinea. Courtesy of D. Northcutt. (See Plate 7.28 b for habitat.)

Plate 7.38. *Varanus primordius*, Arnhem Land, Northern Territory, Australia. Photo by K. H. Switak.

Plate 7.39 a. *Varanus rosenbergi*, West Head, New South Wales, Australia. Courtesy of R. Hoser.

Plate 7.39 b. Habitat of *V. rosenbergi*, vicinity of Coffin Bay, South Australia. Photo by author.

Plate 7.40. *Varanus rudicollis*, Kalimantan (Borneo), Indonesia. Photo by W. B. Love.

Plate 7.41. *Varanus salvadorii*, Irian Jaya, New Guinea. Courtesy of D. Northcutt.

Plate 7.42. *Varanus salvator salvator*, Malaya. Courtesy of D. Northcutt.

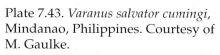

Plate 7.43. *Varanus salvator cumingi*, Mindanao, Philippines. Courtesy of M. Gaulke.

Plate 7.44. *Varanus salvator marmoratus*, Sibutu Island, Philippines. Courtesy of M. Gaulke.

Plate 7.45. *Varanus salvator nuchalis*, Cebu Island, Philippines. Courtesy of M. Gaulke.

Plate 7.46 a. *Varanus scalaris*, Darwin, Northern Territory, Australia. Courtesy of R. W. Van Devender.

Plate 7.46 b. Habitat of *V. scalaris*, *V. storri*, and *V. tristis*, vicinity of Cloncurry, Queensland, Australia. Courtesy of R. W. Van Devender.

Plate 7.47 a. *Varanus spenceri*, Winton, Queensland, Australia. Courtesy of R. W. Van Devender.

Plate 7.47 b. Habitat of *V. spenceri*, vicinity of Camooweal, Northern Territory, Australia. Courtesy of R. Hoser.

Plate 7.48. *Varanus spinulosis*, Santa Isabel Island, Solomons. Photo by author.

Plate 7.49. *Varanus storri storri*, Charters Towers, Queensland, Australia. Courtesy of R. Hoser. (See Plate 7.46 b for habitat.)

Plate 7.50. *Varanus teriae*, York Peninsula, Queensland, Australia. Courtesy of S. T. Irwin.

Plate 7.51. *Varanus timorensis timorensis*, Timor Island, Indonesia. Courtesy of R. W. Van Devender.

Plate 7.52. *Varanus timorensis similis*, Irian Jaya, New Guinea. Courtesy of D. Northcutt.

Plate 7.53. *Varanus tristis orientalis*, Kingooya, South Australia. Courtesy of R. Hoser.

Plate 7.54 a. *Varanus varius*, New South Wales, Australia. Courtesy of R. W. Van Devender.

Plate 7.54 b. *Varanus varius* (banded phase), Angathella, Queensland, Australia. Courtesy of R. W. Van Devender.

Plate 7.54 c. Habitat of *V. varius*, Kurringai Chase, New South Wales, Australia. Courtesy of R. Hoser.

Plate 7.55. *Varanus yemenensis*, Al Kobor, Yemen. Courtesy of W. Böhme.

Chapter 7

SPECIES AND SUBSPECIES OF THE GENUS *VARANUS*

GENUS *VARANUS*

Lacerta Linnaeus, Syst. Nat., 1758, p. 201. (type *Lacerta monitor*)

Tupinambis Daudin, Hist. Nat. Gen. Part. Rept., 1802, p. 20 (type *Tupinambis monitor*)

Monitor (not in Blainville 1816) Lichenstein, Zool. Mus. Univ. Berlin, ed. 2, 181, p. 66 (type *niloticus*).

Varanus Merrem, Tentamen systematis amphibiorum. Marburg, 1820, p. 58 (type *Lacerta varia* Shaw).

Varanus is a latinization of "waran," the Egyptian name for the Nile monitor. The term literally means "monitor" from an ancient belief that these lizards would alert people to the presence of crocodiles . . .

Thus would begin a systematic review of the varanids. In this chapter I will address the taxonomy of each taxon in the genus as currently recognized (1995). A good deal of subjective judgment must be used by any author who ventures to make a list of species of any kind of reptile. Taxonomic standings change as more and different evidence come to light through research. The taxonomy adopted in this chapter incorporates the findings of Wolfgang Böhme, a leading varanid systematist. Several of these have not yet been adopted by the herpetological community as a whole, but seem to be valid in my opinion. Other taxa are the subject of two-sided debates in regard to their status. For these I have had to make a judgment call. Thus I accept full responsibility for the taxonomy adopted. In each account I summarize the history of changes in scientific names (an abbreviated synonymy), in order to assist the reader to reference accounts using an older taxonomy. Where an English common name is widely used for a particular taxon, that is also given.

Each taxon (species or subspecies) account contains a brief description to help make identification possible for those using this chapter as a field guide. Similarly, each account contains both a written version of the distribution and a range map. Each account ends with any special notes on the life history and ecology that have not been included in previous chapters on these subjects. Sizes given are the ranges for adults. Range maps are at the end of this chapter.

Varanus acanthurus acanthurus Boulenger 1885
Northwestern Ridge-tailed Monitor

1845 *Odatria ocellata* Gray, Cat. Liz. Brit. Mus., 1:8

1885 *Varanus acanthurus* Boulenger, Cat. Liz. Brit. Mus. 2:324

Description: Dorsum with reticulated pattern of rich dark brown with numerous small yellow spots, lines, or cream ocelli. Head brown, spotted with yellow or cream, tending to longitudinal stripes on the neck. Dark brown stripe along the snout and through the eye. Tail dark brown with yellow or light brown rings. Whitish below. Head scales small and smooth. Nostril half way between eye and tip of snout, oval. Tail almost round in section with very spiny scales on top and sides. Total length: to 700 mm.
Distribution: Arid northwestern Australia. (Map 7.1)
Ecology: Inhabits rocky outcrops, living under boulders or in deep crevices. Feeds mostly on insects (orthopterans, beetles, cockroaches) and other lizards.

Varanus acanthurus brachyurus Sternfeld 1919
Common Ridge-tailed Monitor

1919 *Varanus acanthurus brachyurus* Sternfeld, Senckenb. 1:78

Description: Similar to the nominal subspecies except that the stripe from the snout through the eye is yellowish. Longitudinal stripes on the back of neck are very prominent. Total length: to 780 mm. (Plate 7.1)
Distribution: Arid and seasonally dry regions of Western Australia, Northern Territory, northern South Australia, and western Queensland. (Map 7.1)
Ecology: Inhabits rocky outcrops and spinifex (*Triodia*) tussocks. Feeds on insects and small lizards.

Varanus acanthurus insulanicus Mertens 1958
Island Ridge-tailed Monitor

1958 *Varanus acanthurus insulanicus* Mertens, Senckenb. 39:229

Description: Similar to *V. a. brachyurus*, but larger (SV length over 230 mm) and darker in color (melanistic).
Distribution: Groote Eylandt and Marchinbar Is., in the Gulf of Carpentaria. (Map 7.1)

Varanus albigularis albigularis (Daudin 1802)
South African White-throated Monitor

1802 *Tupinambis albigularis* Daudin, Hist. Nat. Rept. 3:72

1831 *Monitor albigularis* Gray, Griff. Anim. Kingd. 9:28

1895 *Varanus albigularis* Boulenger, Cat. Liz. Brit. Mus. 2:307

1919 *Varanus exanthematicus albigularis* Schmidt, Bull. Amer. Mus. Nat. Hist. 39:483

1988 *Varanus albigularis albigularis* Böhme, Bonn. Zool. Monog. 27:1

Description: Dark, grayish brown above, with large rounded yellow-white black-edged dorsal spots arranged more or less in transverse zones. Dark temporal line from eye along neck to shoulder. Tail banded alternately with dark brown and yellowish white. Total length: to 2000 mm. Finally recognized as distinct from *V. exanthematicus* by totally different hemipenis morphology (Böhme, 1988). (Plate 7.2)

Distribution: Namibia, Botswana, South Africa, Zimbabwe, Mozambique, Zambia. (Map 7.3)

Ecology: Inhabits savannah and open bush country. Diet extremely varied, including toads and carrion, especially fond of large beetles.

Varanus albigularis angolensis Schmidt 1933
Angola White-throated Monitor

1933 *Varanus albigularis angolensis* Schmidt, Ann. Carnegie Mus. 22:10

1937 *Varanus exanthematicus angolensis* Mertens, Abh. Senckenb. Nat. Ges. 435:9

1988 *Varanus albigularis angolensis* Böhme, Bonn. Zool. Monog. 27:1

Description: Similar to the nominal subspecies, but with larger body scales. Total length: to 1125 mm.

Distribution: Angola. (Map 7.3)

Varanus albigularis ionidesi Laurent 1964
Ionides' White-throated Monitor

1964 *Varanus exanthematicus ionidesi* Laurent, Breviora 199:1-5.

1988 *Varanus albigularis ionidesi* Böhme, Bonn. Zool. Monog. 27:1

Description: Adults resemble *V. a. microstictus* and juveniles resemble young *V. a. albigularis*. The general color is lighter and with less contrast than *microstictus*. Two very distinct black lines run from the eye to the neck where they converge. (Plate 7.3)

Distribution: southeastern Tanzania. (Map 7.3)

Varanus albigularis microstictus Boettger 1893
East African White-throated Monitor

1893 *Varanus microstictus* Boettger, Kat. Rept.-Samml. Mus. Senckenb. Nat. Ges. 1:72

1942 *Varanus exanthematicus microstictus* Mertens, Abh. Senckenb. Nat. Ges. 466:355

1988 *Varanus albigularis microstictus* Böhme, Bonn. Zool. Monog. 27:1

Description: Similar to nominal subspecies, except the dorsal spots are whitish without the dark edges, arranged in transverse rows. Dark temporal stripe present does not extend to shoulder. Total length: to 2000 mm. (Plate 7.4)

Distribution: southern Ethiopia, southern Somalia, Kenya, Tanzania, Uganda, Mozambique. (Map 7.3)

Varanus baritji King and Horner, 1987
White's Monitor

1987 *Varanus baritji* King and Horner, The Beagle 4:73

Description: Similar to *V. acanthurus* except in color pattern. Reddish brown above with numerous scattered small dark spots; head usually plain brown; distinctive lemon yellow throat. Total length: to 720 mm.

Distribution: northern Northern Territory, Australia. (Map 7.1)

Ecology: found in rocky limestone and sandstone outcrops, rocky hills and slopes with compacting loam soils, within tropical woodlands and shrublands with grassy understory.

Varanus beccarii (Doria 1874)
Black Tree Monitor

1874 *Monitor beccarii* Doria, Ann. Mus. Civ. Stor. Nat. Genova, 6:331.

1941 *Varanus prasinus beccarii* Mertens, Abh. Senckenb. Nat. Ges, 23:272.

1991 *Varanus beccarii* Sprackland, Mem. Queensland Mus. 30:572

Description: Jet black throughout. Head scales small and smooth; nuchal scales keeled. Nostril on anterior canthus of eye, close to snout, oval. Tail compressed laterally, twice as long as head and body, tip usually kept coiled at rest. Caudal scales strongly keeled, in rings. Total length: to 900 mm. Insular melanism has occurred several times in the radiation of the genus *Varanus*. (Plate 7.5)

Distribution: Aru Islands, Indonesia. (Map 7.4)

Ecology: Arboreal, but readily takes to water. Feeds chiefly on geckos and frogs.

Varanus bengalensis bengalensis (Daudin 1802)
Bengal Monitor

1758 *Lacerta monitor* Linnaeus, Syst. Nat. 10:201

1802 *Tupinambis bengalensis* Daudin, Hist. Nat. Rept. 3:67

1885 *Varanus bengalensis* Boulenger, Cat. Liz. Brit. Mus. 2:310.

Description: Adults brownish or olive above, usually with blackish dots; lower parts yellowish, uniform or spotted with black. Juveniles olive spotted with numerous yellow spots or ocelli often arranged in transverse bars; underparts whitish with narrow dark transverse bars. Snout convex; nostril an oblique slit, closer to eye than tip of snout. Scales on top of head larger than nuchal scales. Tail strongly compressed, with short double-toothed crest. Total length: to 2000 mm. (Plate 7.6)

Distribution: southeastern Iran, Afghanistan, Sri Lanka, Pakistan, Bangladesh, western Burma, northern India, Nepal. (Map 7.5)

Ecology: Scrub forest and other areas with a pronounced dry season, to the lower edge of pine-oak forest in Himalayas. A good runner, this lizard usually climbs trees to escape enemies, taking refuge in hollows. Can drop to the ground from heights of 10 m without being injured. Food: small mammals, frogs, beetles, lizards, grasshoppers.

Varanus bengalensis irrawadicus Yang and Li 1987
Yunnan Monitor

1987 *Varanus irrawadicus* Yang and Li, Chinese Herp. Res. 1:60-63.

1996 *Varanus bengalensis irrawadicus* De Lisle, Nat. Hist. Monitor Lizards :119.

Status: Auffenberg (1994) claims this form is identical to the nominal form. Yang Datong and Liu Wanzhao (1994) claim that it varies from the nominal race in 4 of 15 characters examined, and lists it as a separate species. Its range is not contiguous with that of the nominal race and most closely approaches that of *V. b. nebulosus*. I believe this to be a geographically isolated population distinct enough to warrant subspecific status, but no specific status at this time.

Description: Similar to *V. b. bengalensis* except supraoculars small and not widened; nostril midway between eye and snout, slit-like; chevron-shaped marking on nape absent. Back blackish brown with small yellow spots; marbling on and under neck; head with black reticular markings; temporal stripe dark and distinct; tail bars indistinct or absent. Total length: to 1240 mm. (Plate 7.7)

Distribution: Known from the Wanding Valley, Yunnan, China, and adjacent northwestern Vietnam. (Map 7.5)

Varanus bengalensis nebulosus (Gray 1851)
Clouded Monitor

1831 *Monitor nebulosus* Gray, Griff. Anim. King. 9:27(suppl.)

1836 *Varanus nebulosus* Dumeril and Bibron, Esp. Gen. 3:483

1942 *Varanus bengalensis nebulosus* Mertens, Abh. Senckenb. Nat. Ges. 466:332

Description: Dark olive or brown above, with numerous yellow spots; chin and throat with transverse black bands. Top of head yellow. Lower parts marbled yellow and dark brown. Young with yellow ocelli on the back arranged in transverse series. Snout convex; nostril oblique slit, nearer eye than tip of snout. Scales on top of head smooth and larger than nuchal scales; scales on back strongly keeled. Tail strongly compressed with a low double-toothed crest. Total length: to 1430 mm. (Plate 7.8)
Distribution: southern Burma, Thailand, Malaysia (western), Indonesia (Java), Vietnam. (Map 7.5)
Ecology: Usually found in open tropical deciduous forest, often in the vicinity of villages. Lashes out violently with tail when cornered.

Varanus bengalensis vietnamensis Yang and Liu 1994
Vietnam Monitor

1994 *Varanus vietnamensis* Yang and Liu, Chinese Zool. Res. 15

1996 *Varanus bengalensis vietnamensis* De Lisle, Nat. Hist. Monitor Lizards :120

Status: Recently described form from northern Vietnam. I judge it to resemble *V. bengalensis* too closely to warrant specific status at this time.
Description: Resembles *V. b. bengalensis* except for widened supraocular scales, naris is closer to the snout than the eye, scales on the crown are smaller than the nuchal scales. It lacks any of the transverse banding of adjacent populations of *V. b. nebulosus*. (Plate 7.9)
Distribution: Known only from the highlands of north Vietnam. (Map 7.5)

Varanus bogerti Mertens 1950
Bogert's Tree Monitor

1950 *Varanus prasinus bogerti* Mertens, Amer. Mus. Novit. 1456:1–7.

1991 *Varanus bogerti* Sprackland, Mem. Queensland Mus. 30:571

Description: Similar to *V. prasinus*, but the color is coal black; head scales rugose; nuchal scales tuberculated. Differs from *V. beccarii* by lower scale counts. Total length: to 830 mm
Distribution: d'Entrecasteaux and Trobriand Archipelagos, Papua New Guinea. (Map 7.4)

Varanus brevicauda Boulenger 1898
Short-tailed Monitor

1898 *Varanus brevicauda* Boulenger, Proc. Zool. Soc. London 56: 916,920

Description: Head and body pale yellowish to pinkish brown with numerous small spots of dark brown or cream. White below. Head scales small, smooth. Nostril lateral, half way between eye and tip of snout, oval. Tail roundish, slightly shorter than head and body (the only monitor with a tail shorter than S–V length). Total length: to 230 mm (smallest species). (Plate 7.10)

Distribution: Arid regions of northern Western Australia, southern Northern Territory, western Queensland. (Map 7.2)

Ecology: a terrestrial species of the sandy desert. Inhabits spinifex (*Triodia*) tussocks or litter; also compacting loam with surface rocks. Hibernates during coldest time of year (June–July). Feeds on reptile eggs and insects (grasshoppers, beetles, roaches, caterpillars). Mates in spring, eggs laid in September–October.

Varanus caudolineatus Boulenger 1885
Stripe-tailed Monitor

1885 *Varanus caudolineatus* Boulenger, Cat. Liz. Brit. Mus. 2:324.

Description: Gray brown above with numerous dark brown spots. Tail gray with a series of dark brown longitudinal stripes. Dark brown stripe on head. White below, with some brown spots. Head scales small, smooth. Nostril lateral, half-way between eye and tip of snout, oval. Tail roundish; scales spiny. Total length: to 320 mm. (Plate 7.11)

Distribution: Coast and interior of central Western Australia. (Map 7.6)

Ecology: Mostly arboreal, living under bark or in holes of mulga (*Acacia*) and gum (*Eucalyptus*) trees. Sometimes saxicolous, in exfoliating granite outcrops. Feeds on insects (chiefly grasshoppers, roaches) and small lizards (chiefly geckos). Mating appears to occur in July and August (mid-winter), although the 3-5 eggs are not laid until November or December.

Varanus doreanus doreanus (A.B. Meyer 1874)
Blue-tailed Monitor

1830 *Monitor kalabeck* Lesson, Duperrey's Voyage Coquille Zool. 2:52

1874 *Monitor doreanus* A.B. Meyer, Mber. Akad. Wiss. Berlin 1874:130

1878 *Varanus kalabeck* Peters & Doria, Ann. Mus. Civ. Stor. Nat. Genova 13:330

1942 *Varanus indicus kalabeck* Mertens, Abh. Senckenb. Nat. Ges. 466:270

1994 *Varanus doreanus doreanus* Böhme, Salaman. 15:134

Description: Similarity of appearance of preserved specimens caused Mertens to consider this a subspecies of *V. indicus*. Neck scales larger than *indicus*. Head is dark brown; spots bright yellow inside paler yellow-edged polygons; tail with sky blue reticulation; feet bluish; neck dark with large whitish spots. Total length: to 1200 mm. (Plate 7.12)

Distribution: Known from Halmahera Island (Moluccas); Waigeo Island (west of New Guinea); Sorong, Vogelkop, Jayapura, and Merauke areas of Irian Jaya, Indonesia; Madan, Papua New Guinea. (Map 7.28)

Varanus doreanus finschi Bohme 1994
Finsch's Monitor

1994 *Varanus doreanus finschi* Böhme, Salaman. 15:137

Description: Similar to the nominal subspecies, but with neck mostly white; higher scale count on mid-back; smaller. Total length: to 400 mm.
Distribution: Known only from New Britain Island, Bismarck Archipelago. (Map 7.28)

Varanus dumerilii dumerilii (Schlegel 1839)
Dumeril's Monitor

1839 *Monitor dumerilii* Schlegel, Abb. Amphib. p. 78

1885 *Varanus dumerilii* Boulenger, Cat. Liz. Brit. Mus. 2:312.

Description: Brown above, spotted with black with four narrow pale yellow cross-bars on back; limbs dark brown, spotted with yellow. Snout tip depressed; nostril an oblique slit close to eye. Nuchal scales very large, posterior ones keeled. Scales on back large, irregular in size, keeled. Tail strongly compressed, with a low double-toothed crest. Total length: to 1250 mm. (Plate 7.13)
Distribution: southern Thailand, Malaya, Indonesia (Sumatra, Banka, Kalimantan, Riou, and Billiton). (Map 7.7)
Ecology: Semiarboreal, inhabiting evergreen forests and mangrove swamps. Feeds on crabs and insects. Mating occurs in July–August; eggs laid in September–October.

Varanus dumerilii heteropholis Boulenger 1892
Sarawak Forest Monitor

1892 *Varanus heteropholis* Boulenger, Proc. Zool. Soc. London 29:506

1942 *Varanus dumerilii heteropholis* Mertens, Abh. Senckenb. Nat. Ges. 466:366

Description: Similar to the nominal subspecies. Tail less compressed; neck and body scales larger. Total length: to 1050 mm.
Distribution: Sarawak (Map 7.7).
Ecology: Inhabits primary and secondary forests.

Varanus eremius Lucas and Frost 1895
Rusty Desert Monitor

1895 *Varanus eremius* Lucas and Frost, Proc. Zool. Soc. Victoria 7:267

Description: Reddish brown above, with numerous dark brown and pale yellow spots. Tail with alternating stripes of cream and dark brown. Black stripe from snout to eye. White below; throat marked with dark symmetrical blotches. Head scales small, keeled. Nostril lateral, near tip of snout, oval. Tail roundish; caudal scales keeled but not spiny. Total length: to 500 mm. (Plate 7.14)

Distribution: Central and coastal Western Australia, deserts of South Australia and Northern Territory. (Map 7.8)

Ecology: Terrestrial; sandy and loamy deserts, in and around spinifex (*Triodia*) tussocks. Does not hibernate. Forages widely in search of insects (chiefly grasshoppers) and small lizards. Mating occurs September to November. Eggs are laid in January or February, hatching in March or April.

Varanus exanthematicus (Bosc 1792)
Savannah Monitor

1792 *Lacerta exanthematicus* Bosc, Act. Soc. Hist. Nat. Paris 1:25

1820 *Varanus exanthematicus* Merrem, Tent. Syst. Amphib. p. 60

Description: Grayish brown above, with yellowish, black-edged spots; limbs spotted with yellowish white; dirty yellowish white below. Snout rounded, convex; nostril an oblique slit close to eye. Scales on head small, flat, granular. Scales on tail slightly keeled; tail roundish in section with low toothed crest. Toes short. Body stout in appearance. Total length: to 1000 mm. (Plate 7.15)

Distribution: Senegal east to Sudan, Eritrea and Ethiopia. (Map 7.3)

Ecology: Terrestrial, in open savannah. It is, however, an excellent climber both of trees and rocks. Occupies rock crevices, holes in trees, and abandoned small mammal burrows. The species is not a fast runner, and when cornered it arches the neck, hisses, and lashes out with the tail. If all else fails, it will feign death. Feeds on small mammals, birds and their eggs, other reptiles, frogs, snails, scorpions, and insects. Mating occurs August–October; eggs laid November–January.

Varanus flavescens (Hardwicke and Gray 1827)
Yellow Monitor

1827 *Monitor flavescens* Hardwicke and Gray, Zool. Jour. 3:226

1864 *Varanus flavescens* Guenther, Rept. Brit. Ind. p. 65

Description: Back and tail have pattern of alternating transverse bars of reddish brown and dirty yellow. Ventral area yellow. Dark temporal stripe. During the breeding season orange-red suffusion spreads over the body; only varanid species known to undergo seasonal color change. Snout short, convex. Nostril an oblique slit near tip of snout. Tail strongly compressed, with a low double-toothed crest. Total length: to 1000 mm. Only species known in which males are not larger than females. (Plate 7.16)

Distribution: Flood plains of Indus, Ganges, and Brahmaputra rivers of Pakistan, India, Nepal, and Bangladesh. (Map 7.9)

Ecology: Inhabits grassy and shrubby wetlands. Feeds on insects and worms. Eggs laid in August and September.

Varanus flavirufus flavirufus Mertens 1958
Sand Monitor

1838 *Hydrosaurus gouldii* Gray, Ann. Nat Hist. 1:394

1851 *Varanus gouldii* Dumeril, Cat. Meth. Coll. Rept. Mus. Paris :52

1958 *Varanus gouldii flavirufus* Mertens, Senckenb. Biol. 39:229

1991 *Varanus flavirufus flavirufus* Bohme, Mertensiella 2:38

Description: Dorsal ground color reddish brown with numerous small light spots usually aligned transversely to form cross-bands. Well defined black temporal stripe edged with white. Tail scales form rings; tip of tail bright yellow. Whitish below. Head scales small, irregular, smooth. Nostril lateral, nearer tip of snout, oval. Tail strongly compressed, with distinct double-toothed crest. Total length: to 1400 mm. (Plate 7.17)

Distribution: Arid interior of Australia and northern region. (Map 7.10)

Ecology: Inhabits the sandy desert regions, sheltering in burrows at night and in the heat of the day.

Varanus flavirufus gouldii (Gray 1838)
Bungarra

1838 *Hydrosaurus gouldii* Gray, Ann. Nat Hist. 1:394

1851 *Varanus gouldii* Dumeril, Cat. Meth. Coll. Rept. Mus. Paris p. 52

1958 *Varanus gouldii gouldii* Mertens, Senckenb. biol. 39:229

1991 *Varanus flavirufus gouldii* Bohme, Mertensiella 2:38

Description: Similar to nominal form except larger and darker color. Head and neck dark brown; transverse zones and bands whitish; tip of tail whitish; gray chevron on throat. Total length: to 1600 mm. (Plate 7.18)

Distribution: Widely distributed throughout eastern Australia, except the southeast (Map 7.10)

Ecology: Terrestrial, found in a large variety of habitats, from coastal eucalypt forests to arid scrub lands. Shelters in burrows, hollow logs, or dense litter. Ranges over large areas foraging for insects, frogs, reptiles and their eggs, birds, mammals, and carrion. Large individuals raise themselves on hind legs to obtain a view of surroundings, or when on the defensive. Mating occurs in January; eggs are laid in February.

Varanus giganteus (Gray 1845)
Perentie

1845 *Hydrosaurus giganteus* Gray, Cat. Liz. Brit. Mus. 1:13

1885 *Varanus giganteus,* Boulenger Cat. Liz. Brit. Mus. 2:320

Description: Rich brown above, with cream yellow spots, arranged in transverse rows on body and tail. Light spots edged with dark brown or black which run together to form reticulated pattern on sides of neck and face. Whitish below. Head scales small, smooth irregular. Nostril lateral, close to tip of snout, oval. Tail strongly compressed, with distinct crest on posterior half. Total length: 1600–2500 mm.; weight to 15 kg. (Plate 7.19)

Distribution: Arid interior of Australia from western Queensland to the coast of Western Australia. (Map 7.11)

Ecology: Terrestrial, inhabiting deep crevices and burrows in rocky outcrops, but forages widely on open desert. Sometimes climbs trees if disturbed. Feeds on insects, reptiles, birds, small mammals, and carrion. Mating occurs in spring, eggs laid in January. Young emerge the following spring.

Varanus gilleni Lucas and Frost 1895
Pygmy Mulga Monitor

1895 *Varanus gilleni* Lucas & Frost, Proc. Soc. Victoria 7:266

Description: Rich gray brown above, gray on sides, with a series of narrow dark reddish brown cross-bands. Tail gray with dark brown cross-bands anteriorly, and longitudinal stripes posteriorly. White below with small gray spots. Head scales small, smooth, irregular. Nostril lateral, half way between eye and tip of snout, oval. Tail roundish. Total length: to 340 mm. (Plate 7.20)

Distribution: Desert areas of South Australia and Northern Territory, through the interior of Western Australia to the northwest coast. (Map 7.12)

Ecology: Arboreal, usually under bark or in hollows of mulga (*Acacia*), gum (*Eucalyptus*), or desert oak trees (*Casuarina*). Feeds on insects (othopterans, beetles), spiders, bird eggs, and small lizards. Mating occurs in September and October.

Varanus glauerti Mertens 1957
Glauert's Monitor

1957 *Varanus timorensis glauerti* Mertens, West. Aust. Nat. 5:183.

1958 *Varanus glauerti* Mertens, Sencken. Biol. 39:229–264.

Description: Blackish or dark brown above, with obscure transverse rows of large light spots across back. Distinct white temporal stripe. Tail very long with alternate light and dark cross-bands. Throat and belly whitish. Head scales small, smooth, irregular. Nostril lateral, nearer tip of snout than eye, oval. Tail roundish. Total length: to 800 mm. (Plate 7.21)

Distribution: Known from the Kimberley region of Western Australia and extreme northwestern Northern Territory. (Map 7.12)

Ecology: Rock outcrops, boulder fields, walls of gorges in tropical and riverine forest. Feeds on orthopterans, spiders, and small lizards, reptile eggs.

Varanus glebopalma Mitchell 1955
Long-tailed Rock Monitor

1957 *Varanus glebopalma* Mitchell, Rec. S. Aust. Mus. 11:389.

Description: Black above with small fawn-colored spots which form a reticulum over the flanks. Posterior of tail creamy yellow. Throat white with purplish brown reticulum; belly and chest white with transverse bars of light purplish fawn. Head scales small, irregular, smooth. Nostril lateral, near tip of snout, oval. Tail roundish. Total length: to 1060 mm. (Plate 7.22)

Distribution: northeast Western Australia, northern Northern Territory, to northwest Queensland. (Map 7.11)

Ecology: Rugged tropical sandstone and quartzite hills, especially vicinity of gorges or escarpments. Shelters in crevices, caves, and under large boulders. Commonly crepuscular, foraging just after sundown. Lives in colonies. Feeds on lizards, frogs, and insects, chiefly orthopterans.

Varanus gouldii gouldii (Gray 1838)
Gould's Monitor

1838 *Hydrosaurus gouldii* Gray, Ann. Nat. Hist. 1:394

1839 *Monitor gouldii* Schlegl, Abb. Amphib. p. 78

1851 *Varanus gouldii* Dumeril and Dumeril, Cat Meth. Coll. Rept. Mus. Paris p. 52

1980 *Varanus panoptes* Storr, Rec. West. Aust. Mus. 8:273–276.

1991 *Varanus gouldii gouldii* Böhme, Mertensiella 2:38

Description: Dark brown above, usually with pattern of alternating transverse rows of pale spots and dark spots. Limbs with numerous pale spots. White-edged, dark brown temporal stripe. End of tail pale with dark bands. Whitish below with transverse rows of small dark spots. Head scales small, irregular, smooth. Nostril lateral, nearer tip of snout than eye, oval. Tail strongly compressed with double-toothed crest. Total length: to 1600 mm. (Plate 7.23)

Distribution: Kimberley region of northeastern Western Australia and adjacent Northern Territory. (Map 7.13)

Ecology: Terrestrial on temperate to tropical riverside woodland, lagoons, and black-soil flood plains. Feeds on insects (orthopterans, ants, lepidopterans), frogs, lizards, and small mammals.

Varanus gouldii horni Böhme 1988
Horn's Monitor

>1988 *Varanus panoptes horni* Böhme, Salamandra 24:87.
>
>1991 *Varanus gouldii horni* Böhme, Mertensiella 2:38

Description: Similar to the nominal subspecies, but grayer ground color. (Plate 7.24)
Distribution: semiarid region of southern New Guinea. (Map 7.13)

Varanus gouldii rubidus Storr 1980
Yellow-spotted Monitor

>1980 *Varanus panoptes rubidus* Storr, Rec. West. Aust. Mus. 8:276–277
>
>1991 *Varanus gouldii rubidus* Böhme, Mertensiella 2:38

Description: Similar to the nominal subspecies, except dorsal ground color is reddish brown. Total length: to 700 mm.
Distribution: southern Pilbara south to Fields Find and Mt. Linden, Western Australia. (Map 7.13)
Ecology: Terrestrial on read loamy soil with *Acacia* cover and spinifex grass understory.

Varanus griseus griseus (Daudin 1803)
Western Desert Monitor

>1803 *Tupinambis griseus* (Daudin, Hist. Nat. Rept. 8:352)
>
>1820 *Varanus scincus* Merrem, Tent. Syst. Amphib. p. 59
>
>1885 *Varanus griseus* Boulenger, Cat. Liz. Brit. Mus. 2:306

Description: Grayish brown to yellowish brown above, with small yellowish spots; brown cross-bars on back and tail, and two or three longitudinal brown lines on neck; yellowish below. In juveniles dorsal bars are almost black and yellow spots are much larger. Snout depressed. Nostril an oblique slit, closer to eye than tip of snout. Scales on top of head larger than nuchal scales. Tail round. Total length: to 1500 mm. (Plate 7.25)
Distribution: Western Sahara and Mauritania east to Sudan and Egypt in North Africa; Syria, Lebanon, Israel, Jordan, Iraq and the Arabian Peninsula in Asia. (Map 7.14)
Ecology: Inhabits sandy desert, usually where the ground is undulating and vegetation sparse. A good runner. Sometimes digs its own burrow, but more often adopts abandoned mammal burrows to which it retreats during

the heat of the day. Feeds on small rodents, lizards, snakes, scorpions, insects, and carrion. Courtship season (in Israel) May–June, eggs are buried in sand or placed at the end of the burrow.

Varanus griseus caspius (Eichwald 1831)
Eastern Desert Monitor

1831 *Psammosaurus caspius* Eichwald, Zool. Spec. 3:190

1845 *Varanus caspius* Gray, Cat. Liz. Brit. Mus. 1:7

1963 *Varanus griseus caspius* Anderson, Proc. Calif. Acad. Sci. (4)31:417–498

Description: Similar to the nominal subspecies, with dorsal spots more brown than yellow. Total length to 1,325 mm. (Plate 7.26)
Distribution: Iran, Afghanistan, northern Pakistan, Turkmenistan, so. Kazakhstan, Uzbekistan, Tadzhikistan. (Map 7.14)

Varanus griseus koniecznyi Mertens 1954
(Thar Desert Monitor)

1869 *Varanus ornatus* Carlleyle, Jour. Asiat. Soc. Bengal 38:192

1954 *Varanus griseus koniecznyi* Mertens, Senck. Biol. 35:353–357

Description: Dorsal head color gray to black; body dull yellow to light gray with 3 to 5 dark gray cross bands on the body, each bordered on both sides by a row of pale spots; tail with 8 to 15 dark bands, last part distinctly keeled. Dark stripe extending from canthus through the ear to neck. Total length: to 750 mm. (Plate 7.27)
Distribution: western Pakistan to northwest India. (Map 7.14)
Ecology: Favors arid desert and steppe with a sandy substrate, in some places with rocky hills. Most eggs are laid in September, and young hatch the following July to September during the monsoon.

Varanus indicus (Daudin 1802)
Mangrove Monitor

1802 *Tupinambis indicus* Daudin, Hist. Nat. Rept. 3:46

1876 *Monitor indicus* Peters, Mber. Akad. Wiss. Berlin p. 530

1883 *Varanus indicus* Boulenger, Proc. Zool. Soc. London 1883:386

1926 *Varanus indicus indicus* Mertens, Senckenb. 8:274

1994 *Varanus indicus* Böhme, Salaman. 15:133

Description: Dark purplish brown to black above, with numerous spots of cream, yellow, or yellow green. Whitish below. Head scales small and smooth except above the eye. Nostril lateral, slightly nearer tip of snout than eye, oval. Tail strongly compressed, with a low dorsal crest. Total length: to 1400 mm. (Plate 7.28)

Distribution: Sulawesi, Moluccas, Aru Islands, Talaud, Irian Jaya, and Timor in Indonesia; eastern Cape York Peninsula, the islands of Torres Strait, and coastal Arnhem Land in Australia; Solomon Islands; Papua New Guinea and the Bismarck Archipelago; Caroline Islands; Mariana Islands; Marshall Islands. (Map 7.15)

Ecology: Arboreal and aquatic, inhabiting rain forest and coastal mangroves. Excellent swimmer and climber, it forages for insects, crabs, fish, reptiles and their eggs, birds and their eggs, and small mammals in and near forest streams and tidal mangrove areas. Usually dives into the water when disturbed. Nests in rotting wood.

Varanus jobiensis Ahl 1932
Peach-throated Monitor

1932 *Varanus indicus jobiensis* Ahl, Mitt. Zool. Mus. Berlin 17:892

1951 *Varanus karlschmidti* Mertens, Field. Zool. 31:467–471

1991 *Varanus jobiensis* Bohme, Mertensiella 2:46

Description: Blackish above with numerous yellow or cream-colored spots. Whitish below, but with a pinkish yellow throat. Nostril lateral, closer to tip of snout than eye, oval. Tail laterally compressed. Total length: to 1200 mm. (Plate 7.29)

Distribution: northern New Guinea (including Eastern and Southern Highlands of PNG) from Yapen Is. east to Sepik River valley. (Map 7.29)

Varanus kingorum Storr 1980
Kings' Monitor

1980 *Varanus kingorum* Storr, Rec. West. Aust. Mus. 8:237

Description: Reddish brown above, spotted with black. Whitish below, spotted with dark brown. Head scales small, smooth. Nostril dorso-lateral, halfway between eye and tip of snout, oval. Tail slightly compressed. Total length: to 400 mm. (Plate 7.30)

Distribution: Known only from the east Kimberley region on the northern border of Western Australia and Northern Territory. (Map 7.2)

Ecology: Lives in crevices and under exfoliating sandstone slabs in tropical rocky hills with open woodland and shrubs. Very shy. Feeds on insects (orthopterans, roaches, termites). Sometimes found in colonies.

Varanus komodoensis Ouwens 1912
Komodo Monitor

1912 *Varanus komodoensis* Ouwens, Bull. Jard. Buitenzorg 6:1

Description: Most massive of all varanids (over 50 kg), and largest of living lizards. Body shape is flatter than most species, hind legs stout. Toes have shorter claws than other varanids. Adults uniform gray; females with some reddish on flanks, large males with yellowish green spots on snout. Juveniles are heavily spotted with yellow. Tongue yellow. Nostril oval. Total length: to 3000 mm. (Plate 7.31)

Distribution: Komodo, Padar, Rintja, and Flores Islands, Lesser Sunda region of Indonesia. (Map 7.17)

Ecology: Chief habitat is tropical savannah forest, but ranges widely over the islands, from beach to ridge tops. Seeks refuge in burrows at night in typical varanid fashion, but burrows are barely big enough for the lizard to fit. It is the major predator of the islands, chiefly using the sit-and-wait foraging mode along game trails. Its chief food is deer and wild pig, but it will eat most any kind of animal and can detect carrion from considerable distance and actively seeks it out. Mating occurs July to September; eggs are laid about 30 days later. Largely arboreal until eight months old, feeding on insects and lizards.

Varanus mertensi Glauert 1951
Mertens' Water Monitor

1951 *Varanus mertensi* Glauert, West. Aust. Nat. 3:14.

Description: Rich dark brown to black above, with numerous cream or yellow spots. Lower parts whitish, with gray mottling on throat and blue gray cross bars on chest. Head scales moderate, regular, smooth. Nostrils on top of snout, oval. Tail strongly compressed with a dorsal crest. Total length: to 1170 mm. (Plate 7.32)

Distribution: Coastal and inland waters northeastern Western Australia, northern Northern Territory and northern Queensland. (Map 7.8)

Ecology: Aquatic, basking on rocks or logs in the water, or on tree branches overhanging rivers, swamps, and lagoons. Dives in the water if disturbed, and can remain submerged for long periods. Feeds chiefly on crabs; also eats insects (hemipterans, coleopterans), shrimp, amphipods, fish, frogs, and carrion. Has been observed using its body and tail to herd fish into shallow water. Breeding probably takes place throughout the year.

Varanus mitchelli Mertens 1958
Mitchell's Water Monitor

1958 *Varanus mitchelli* Mertens, Senckenb. Biol. 39:229

Description: Dark brown, blue black, or blackish above with scattered yellow spots or ocelli. Throat and sides of neck bright yellow. Tail black with yellow spots. Ventral surface cream. Head scales moderate size, smooth. Nostril on anterior canthus of eye, closer to tip of snout than to eye, oval. Tail strongly compressed with distinct crest, caudal scales in rings. Total length: to 960 mm. (Plate 7.33)

Distribution: northeastern Western Australia and northern Northern Territory. (Map 7.6)

Ecology: Aquatic, basking on rocks or trees in or beside rivers and lagoons. Feeds on insects (orthopterans), spiders, crabs, fish, and frogs. Breeds in the dry season.

Varanus niloticus (Linnaeus 1758)
Nile Monitor

1758 *Lacerta monitor* Linnaeus, Syst. Nat. 1:201

1818 *Monitor niloticus* Lichtenstein, Zool. Mus. Univ. Berlin 2:66

1826 *Varanus niloticus* Fitzinger, Neue Classif. Rept. p. 50 (includes Mertens (1942) *V. n. niloticus* and *V. n. ornatus*)

Description: Grayish brown to dark olive, with darker reticulation and scattered yellow ocelli on dorsal surface. Lower parts yellowish with grayish cross-bands. Juveniles black above with transverse series of yellow stripes on head, chevron shaped mark on nape. Snout obtusely pointed; nostril round and closer to eye than tip of snout. Scales on head moderate size, polygonal. Tail strongly compressed with a low, toothed dorsal crest. Total length: to 2000 mm. (Plate 7.34)

Distribution: South Africa north to Egypt in the east and Liberia in the west. (Map 7.18)

Ecology: Semiaquatic, always found in the vicinity of permanent rivers or lakes. Excellent swimmer. Feeds on carrion, fish, frogs, crabs, snails, snakes, turtles, eggs of birds and crocodiles, and arthropods. Hatching occurs at beginning of local rainy season; hatchlings feed on insects and small frogs.

Varanus olivaceus Hollowell 1857
Gray's Monitor

1845 *Varanus ornatus* Gray, Cat. Liz. Brit. Mus. 1:10

1857 *Varanus olivaceus* Hollowell, Proc. Acad. Sci. Philad. 1857:10

1885 *Varanus grayi* Boulenger, Cat. Liz. Brit. Mus. 2:312

1988 *Varanus olivaceus* Auffenberg, Gray's monitor lizard p. 3

Description: Olive green above, with transverse double black bands. Gray below. Juveniles forest green above with distinctive black bands. Head yellow. Nostril slit-like, half way from eye to tip of snout. Head scales large. Toes long, with powerful recurved claws. Rear teeth short and blunt. Tail laterally compressed. Total length: to 1500 mm. (Plate 7.35)

Distribution: southern Luzon, Catanduanes Islands, Philippines. (Map 7.19)

Ecology: Arboreal, but forages on the ground. Food spectrum much narrower than any other monitor—fruit and snails. Night spent and escape refuges usually in tree canopies, sometimes in limestone crevices. Mating occurs June through September; eggs are laid about 30 days later, perhaps in tree hollows. Hatching timed to occur during the southwest monsoon. Juveniles feed on insects.

Varanus pilbarenis Storr 1980
Pilbara Monitor

1980 *Varanus pilbarensis* Storr, Rec. West. Aust. Mus. 8:237–293

Description: Pale reddish brown above, head and neck spotted with dark reddish brown. Dark brown ocelli aligned transversely on the back. Tail with narrow reddish brown and gray bands. Whitish below, irregularly spotted or banded with gray. Head scales small, smooth. Nostril facing upwards, halfway between eye and tip of snout, oval. Tail roundish. Total length: to 500 mm. (Plate 7.36)

Distribution: Pilbara region, Western Australia. (Map 7.20)

Ecology: Warm temperate semiarid hillsides and gorges with extensive rock outcrops. Feeds on orthopterans, spiders, and skinks.

Varanus prasinus (Schlegel 1839)
Green Tree Monitor

1839 *Monitor prasinus* Schlegel, Abb. Amphib. 22:78

1856 *Varanus prasinus* Bleeker, Reis Minahassa 1:278

1874 *Monitor kordensis* Meyer, Mber. Berl. Akad. p. 131

1885 *Varanus kordensis* Boulenger, Cta. liz. brit. mus 2:322

1942 *Varanus prasinus prasinus* Mertens, Abh. Senckenb. Nat. Ges. 466:292

1942 *Varanus prasinus kordensis* Mertens, Abh. Senckenb. Nat. Ges. 466:294

1991 *Varanus prasinus* Sprackland, Mem. Queensland Ms. 30:566

Description: Dark jade to lime green above, generally with narrow black transverse cross-bands. Ventral surface pale green. Head scales moderate size, smooth; nuchal scales usually smooth. Nostril on anterior canthus of eye, nearer tip of snout than eye, oval. Tail slightly compressed, twice the length of the body, prehensile and kept coiled when not in use. Total length: to 900 mm. (Plate 7.37)

Distribution: New Guinea and adjacent islands, below 500 m. (Map 7.4)

Ecology: Arboreal, found in monsoon, rain, and palm forests, and in coastal mangroves. Feeds mainly on insects, especially orthopterans (tree crickets); also centipedes and rodents. Lays eggs July–September. Only nest ever found was in a termite-eaten log.

Varanus primordius Mertens 1942
Northern Blunt-spined Monitor

    1942 *Varanus acanthurus primordius* Mertens, Zool. Anz. 138:41

    1966 *Varanus primordius* Storr, Copeia 1966:583–584.

    Description: Reddish brown above, with numerous dark brown scales forming a reticulum. Lower surface whitish. Head scales, small, smooth, irregular. Nostril lateral, slightly nearer tip of snout than eye, oval. Tail round. Total length: to 420 mm. (Plate 7.38)
    Distribution: northern Northern Territory, Australia. (Map 7.2)
    Ecology: Deep cracking tropical alluvial soils and low rocky outcrops. Feeds on insects (ants, orthopterans, roaches), lizards, and reptile eggs.

Varanus rosenbergi Mertens 1957
Rosenberg's Monitor

    1957 *Varanus gouldii rosenbergi* Mertens, Zool. Anz. 159:17.

    1980 *Varanus rosenbergi* Storr, Rec. West. Aust. Mus. 8:280-282

    Description: Blackish above, finely dotted with yellow or white, with about 15 narrow black bands on neck and body. A pale-edged, black temporal stripe. Tail alternately banded blackish brown and pale yellow. Whitish below, reticulated with gray. Head scales small, smooth. Nostril lateral, nearer tip of snout than eye, oval. Tail strongly compressed, with a distinct double-toothed crest. Total length: to 1600 mm. (Plate 7.39)
    Distribution: Far south of Western Australia and Southern Australia, and southwest Victoria. (Map 7.10)
    Ecology: Inhabits cool temperate heath and dry scrub on deep sandy or calcareous soil. Feeds on insects (roaches, orthopterans, beetles), other arthropods (spiders, centipedes, scorpions), birds, small mammals, and other lizards and their eggs. Mating occurs in January; eggs are laid in February, usually in termite mounds.

Varanus rudicollis Gray 1843
Rough-necked Monitor

    1843 *Varanus rudicollis* Gray, Cat. Liz. Brit. Mus. 1:10

    Description: Top of head brown; pair of black stripes begin behind eye and extend to shoulder joined by a median stripe on nape. Sides of neck and under chin reddish; yellowish transverse bands across back in front and behind front legs. Transverse rows of yellowish ocelli across posterior back and tail. Ventral surface dark, with yellowish bars. Melanistic individuals are fairly common. Snout pointed, nostril oblique and slit-like, halfway between eye and tip of snout. Very large prominent keeled scales on nuchal region. Total length: to 1600 mm. (Plate 7.40)

Distribution: southern Burma, southern Thailand, Malaya, Indonesia, (Sumatra, Kalimantan, Banka). (Map 7.21)

Ecology: Arboreal, found only in tropical evergreen forests. It is a highly specialized insectivore, feeding on termites, tree phasmids, and arboreal centipedes; but is also known to take frogs, crabs and spiders.

Varanus salvadorii (Peters and Doria 1878)
Papua Monitor

1878 *Monitor salvadorii* Peters & Doria, Ann. Mus. Civ. Stor. Nat. Genova 13:337

1885 *Varanus salvadorii* Boulenger, Cat. Liz. Brit. Mus. 2:334

Description: Sepia brown above, with numerous yellowish spots over head and neck. Four transverse rows of large round yellow spots on body; spots coalescing to 12 yellow rings on tail. Nostril lateral, close to tip of snout, oval. Total length: to 3000 mm. (Plate 7.41)

Distribution: New Guinea. (Map 7.16)

Ecology: Mostly arboreal, feeding chiefly on birds; but recent work in southern New Guinea shows the species to have consierable terrestrial activity as well.

Varanus salvator salvator (Laurenti 1768)
Asian Water Monitor

1768 *Stellio salvator* Laurenti, Syn. Rept. p. 58

1829 *Monitor nigricans* Cuvier, Regne Anim. 2:27

1834 *Varanus vittatus* Lesson, Voyage Ind. Orient. Zool. p. 307

1845 *Hydrosaurus salvator* Gray, Cat. Liz. Brit. Mus. 1:13

1846 *Monitor salvator* Blyth, J. Asiat. Soc. Bengal 15:376

1847 *Varanus salvator* Cantor J. Asiat. Soc. Bengal 16:635

1932 *Varanus scutigerulus* Barbour, Proc. New Eng. Zoo. club 13:1

1937 *Varanus salvator salvator* Mertens, Senckenb. 19:178 (includes *V. s. scutigerulus* Mertens 1942)

Description: Dark olive above with small yellow spots and large rounded spots or ocelli in transverse rows, becoming indistinct with age. Black temporal stripe. Lower parts yellow with narrow black V-shaped marks on sides of belly. Tail alternately banded with black and whitish. Head long and flat; nostril round, near tip of snout. Scales on top of head larger than nuchal scales. Tail strongly compressed, with double-toothed crest. Total length: to 2500 mm. (Plate 7.42)

Distribution: Sri Lanka, northeastern India, Bangladesh, Burma, Thailand, Kampuchea, Laos, Vietnam, extreme southern China, Malaysia, Indonesia (Sumatra, Nias, Engano, Banka, Kalimantan); the Nicobar Islands. (Map 7.22)

Ecology: Aquatic, frequenting rivers, canals and estuaries. It climbs trees easily in search of food, but never to any great height. When disturbed, it takes to the water, and has been seen swimming far out to sea. Active from early morning to early afternoon, at which time they seek shade under bushes or in trees. Often spends the night in the water. Eggs are laid at the beginning of the rainy season. They are deposited in holes on the banks of rivers or in trees beside the water. Feeds primarily on mammals, small crocodiles, turtles, birds, fish, and crabs.

Varanus salvator adamanensis Deraniyagala 1944
Andaman Islands Water Monitor

1944 *Varanus salvator andamanensis* Deraniyagala, Spol. Zeylanica 24:59–62.

Description: Similar to the nominal form, but darker in overall coloration.

Distribution: Andaman Islands, India. (Map 7.22)

Varanus salvator bivittatus (Kuhl 1820)
Two-striped Water Monitor

1820 *Tupinambis bivittatus* Kuhl, Beit. zur Zool. und Vergleich. Anat., Frankfurt

1835 *Varanus bivittatus* Dumeril and Bibron, Erp. Gen 3:486

Description: Similar to the nominal form, but with brighter yellow markings.

Distribution: Indonesia (Java, Bali, Lombok, Sumbawa, Flores, Wetar). (Map 7.22)

Varanus salvator cumingi Martin 1839
Cuming's Water Monitor

1838 *Varanus cumingi* Martin, Proc. Zool. Soc. Lond. 1838:69

1942 *Varanus salvator cumingi* Mertens, Abh. Senckenb. Nat. Ges. 466:256

Description: Similar to the nominal subspecies, but with larger neck scales. Head yellowish; yellow spots on back coalescing into obscure bands. Tail dark with distinctive yellow cross-bands. Total length: to 1100 mm. (Plate 7.43)

Distribution: Mindanao, Leyte, Samar, Bohol, in the Philippines. (Map 7.22)

Varanus salvator marmoratus (Wiegmann 1834)
Marbled Water Monitor

1834 *Hydrosaurus marmoratus* Wiegman, Reise um die Erde 3:446

1876 *Varanus manilensis* Mertens, Preuss. Exped. Ostas. Zool. 1:196

1888 *Varanus salvator* Gorgoza, Ann. Soc. Esp. Hist. Nat. 17:275

1942 *Varanus salvator marmoratus* Mertens, Abh. Senckenb. Nat. Ges. 466:254

Description: Similar to the nominal subspecies, but with larger neck scales. Ground color light yellowish brown on head, neck, and back, each scale with a dark spot. Yellow spots obscure except on hind legs and tail. Total length: to 1000 mm. (Plate 7.44)

Distribution: Luzon, Palawan, Mindoro, Calamian, in the Philippines; Caroline Islands. (Map 7.15, Map 7.22)

Ecology: Occupies a wide variety of habitats, including mangrove swamps, cultivated fields, and rain forests, from sea level to 1600 m. Feeds on fish, frogs, lizards, birds, rats, crabs, birds' eggs, and carrion.

Varanus salvator nuchalis (Guenther 1872)
Negros Water Monitor

1872 *Hydrosaurus nuchalis* Guenther, Proc. Zool. Soc. Lond. 1872:145

1885 *Varanus nuchalis* Boulenger, Cat. Liz. Brit. Mus. 2:315

1942 *Varanus salvator nuchalis* Mertens, Abh. Senckenb. Nat. Ges. 466:258

Description: Similar to the nominal subspecies with differences in scalation. This race has larger scales on the back, especially the nuchal region. Gray brown above sprinkled with yellow, with a light colored vertebral band. Head brown, becoming yellow in back. Total length: to 1,100 mm. (Plate 7.45)

Distribution: Negros, Guimares, Cebu, Panay, Masbate, Ticao, in the Philippines. (Map 7.22)

Varanus salvator togianus (Peters 1872)
Togian Water Monitor

1872 *Monitor togianus* Peters, Mber. Akad. Wiss. Berlin 1872:582

1885 *Varanus togianus* Boulenger, Cat. Liz. Brit. Mus. 2:316

1941 *Varanus salvator togianus* Mertens, Senckenb. 23:272

Description: Similar to the nominal subspecies but darker; blackish brown with yellow spots on each scale. Total length: to 1100 mm.
Distribution: Togian Islands, Sulawesi, Indonesia. (Map 7.22)

Varanus scalaris Mertens 1941
Australian Spotted Tree Monitor

1941 *Varanus timorensis scalaris* Mertens, Senckenb. 23:266

1983 *Varanus scalaris* Storr, Liz. West. Aust. 2:106

Description: Variable background color: gray to khaki, matching the dominant tree-trunk color of the area; with 10 smokey gray cross-bands across back enclosing yellow spots or ocelli. Distinct black temporal stripe. Total length: to 590 mm. (Plate 7.46)
Distribution: northern part of Australia. (Map 7.24)
Ecology: Open woodland to rainforest. Feeds on insects (orthopterans), scorpions, and lizards.

Varanus semiremex Peters 1869
Rusty Monitor

1869 *Varanus semiremex* Peters, Mber. Akad. Wiss. Berlin 1869:65

Description: Gray brown above with numerous black spots forming a reticulum over the dorsal surface; throat and chest rusty yellow. Whitish below. Head scales small, smooth, irregular. Nostril lateral, nearer tip of snout than eye, oval. Tail round. Total length: to 760 mm.
Distribution: coastal Queensland, Australia. (Map 7.2)
Ecology: Lives in holes in mangrove trees along tidal creeks or around small islands; melaleuca swamps. Feeds on fish, crabs, insects, and lizards (geckos).

Varanus spenceri Lucas and Frost 1903
Spencer's Monitor

1903 *Varanus spenceri* Lucas & Frost, Proc. Soc. Victoria 15:145

Description: Dorsal surface light gray-brown, with scattered dark brown and cream spots. Irregular yellowish cross-bands on neck, belly, and tail. Lower surface cream, spotted with dark gray. Head scales irregular, smooth. Nostril lateral, close to tip of snout, oval. Tail roundish, except the last half which is compressed with a crest. Total length: to 1500 mm. (Plate 7.47)
Distribution: Queensland to eastern Northern Territory, Australia. (Map 7.20)

Ecology: Terrestrial, apparently limited to black-soil plains usually dominated by Mitchell grass, where it shelters in burrows and large cracks in soil. Forages widely, eating orthopterans, isopods, snakes, lizards, small mammals, and carrion. Mates August–October, eggs laid September–November.

Varanus spinulosus Mertens 1941
Solomons Keeled Monitor

1941 *Varanus indicus spinulosus* Mertens, Senckenb. 23:269

1994 *Varanus spinulosis* Sprackland, Herpetofauna 24 (2) :34

Description: Dark brown dorsally with large round dirty yellow spots arranged in four transverse rows across the back; spike-like nuchal and dorsal scales. Total length: to 1000 mm. (Plate 7.48)

Distribution: Known only from Santa Isabel (Bughoto) Is. and adjacent San Jorge Is., Solomons. (Map 7.27)

Ecology: Reported to inhabit forested, mountainous parts of the island.

Varanus storri storri Mertens 1966
Eastern Storr's Monitor

1966 *Varanus storri* Mertens, Senck. Biol. 47:437–441.

Description: Reddish brown above, with numerous dark brown scales forming a reticulated pattern. Dark brown temporal stripe. Undersides pinkish with brown spots. Head scales small, smooth, irregular. Nostril lateral, nearer tip of snout than eye, oval. Tail round; caudal scales spiny. Total length: to 440 mm. (Plate 7.49)

Distribution: northwest Queensland and adjacent Northern Territory, Australia. (Map 7.23)

Ecology: Lives on compacting reddish soils with large surface rocks and tussock grasses. Insectivorous, chiefly orthopterans; also beetles, roaches, ants, and geckos. Hibernates during coldest months (June–July). Burrows into soil under large rocks. Eggs laid in September.

Varanus storri ocreatus Storr 1980
Western Storr's Monitor

1980 *Varanus storri ocreatus* Storr, Rec. West. Aust. Mus. 8:283–285.

Description: Similar to the nominal subspecies. Total length: to 350 mm.
Distribution: northeast Western Australia and adjacent Northern Territory. (Map 7.23)

Varanus telenestes Sprackland 1991
Rossel Island Monitor

> 1980 *Varanus prasinus* Czechura, Mem. Qd. Mus. 20:103
>
> 1991 *Varanus telenestes* Sprackland, Mem. Queensland Mus. 30:569

Description: Similar to *V. prasinus*, but with smooth ventral scales and mottled ventral pattern. Total length: to 425 mm.

Distribution: an island endemic, limited to Rossel I., Papua New Guinea, 350 km from nearest relative (*prasinus*). (Map 7.4)

Varanus teriae Sprackland 1991
Cape York Tree Monitor

> 1856 *Varanus prasinus* Bleeker, Reis Minahassa 1:278
>
> 1980 *Varanus prasinus prasinus* Czechura, Me. Qd. Mus. 20:103
>
> 1991 *Varanus teriae* Sprackland, Mem. Queensland Mus. 30:570

Description: Black above with light bluish green snout, yellow dorsal spots forming thin paired chevrons and caudal rings. Nostril more anterior, body more robust than *V. prasinus*. Lime green ventrally; nuchal scales slightly keeled. Total length: to 515 mm. (Plate 7.50)

Distribution: Nesbit River region, Cape York Peninsula, Australia. (Map 7.4)

Ecology: Found in a variety of habitats, from rain, monsoon, and palm forests to lagoons and mangroves.

Varanus timorensis timorensis Gray 1831
Timor Monitor

> 1831 *Varanus timorensis* Gray, Griff. Anim. Kingd. 9:26
>
> 1836 *Varanus timorensis* Dumeril & Bibron, Esp. Gen 3:473
>
> 1937 *Varanus timorensis timorensis* Mertens, Senckenb. 19:180

Description: Grayish above with black-ringed ocelli covering the back. Limbs spotted white or yellowish. Tail black with rings of whitish scales. Head scales small, smooth. Nostril lateral, a little nearer tip of snout than eye, oval. Tail round. Total length: to 600 mm. (Plate 7.51)

Distribution: Timor, and the small islands off the western end of Timor, Indonesia. (Map 7.24)

Ecology: Arboreal; an active forager for insects both in trees and on ground. Shelters in hollow limbs, holes, or under loose bark. Feeds on scorpions and blind snakes.

Varanus timorensis similis Mertens 1958
New Guinea Spotted Tree Monitor

    1958 *Varanus timorensis similis* Mertens, Senckenb. 39:239

Description: Similar to the nominal subspecies, except ocelli are replaced generally by a series of white or yellowish spots. (Plate 7.52)
Distribution: southern New Guinea. (Map 7.24)

Varanus tristis tristis (Schlegel 1839)
Black-headed Monitor

    1838 *Odatria punctata* Gray, Ann. Nat. Hist. 1:394

    1839 *Monitor tristis* Schlegel, Abb. Amphib. p. 73

    1937 *Varanus timorensis tristis* Mertens, Senckenb. 19:180

    1980 *Varanus tristis tristis* Storr, Rec. West. Aust. Mus. 8:287–292.

Description: Gray to black above, with numerous cream-colored ocelli with dark centers. Head and neck uniformly blackish. Lower surfaces whitish. Head scales smooth, small. Nostril lateral, nearer tip of snout than eye, oval. Tail roundish. Males with spiny scales on each side of vent. Total length: to 800 mm.
Distribution: Western Australia, Northern Territory, and northern South Australia. (Map 7.25)
Ecology: Compacting sandy soils near rocky outcrops, arboreal or saxicolous. Feeds mainly on lizards and orthopterans.

Varanus tristis orientalis Fry 1913
Freckled Monitor

    1913 *Varanus punctatus orientalis* Fry, Rec. Austr. Mus. 10:18

    1942 *Varanus timorensis orientalis* Mertens, Abh. Senckenb. Nat. Ges. 466:301

    1980 *Varanus tristis orientalis* Storr, Rec. West. Aust. Mus. 8:291.

Description: Similar to the nominal subspecies, except the ocelli are larger and fewer. Head and neck resemble body coloration. Total length: to 600 mm. (Plate 7.53)
Distribution: Queensland, northeastern South Australia, northern New South Wales. (Map 7.25)
Ecology: Inhabits tree hollows, rock crevices, or exfoliations in open woodland. Feeds on lizards, grasshoppers, small snakes, and nestling birds.

Varanus varius (White, ex Shaw 1790)
Lace Monitor

> 1790 *Lacerta varia* Shaw, White's Jour. Voyage new South Wales App. p. 246
>
> 1820 *Varanus varius* Merrem, Tent. Syst. Amphib. p. 58

Description: Dark blue black above, with numerous white or yellow scales, spots, or blotches. A banded phase also exists in western part of range. Black bars across snout, chin, and throat. Tail with irregular yellow crossbands. Head scales moderate size, smooth. Nostril lateral, closer to tip of snout than eye, oval. Distinctive ridge of large scales on base of fourth toe. Tail strongly compressed, with distinct double-toothed crest. Total length: to 2100 mm. (Plates 7.54)

Distribution: eastern Queensland, most of New South Wales, southeastern South Australia. (Map 7.23)

Ecology: Arboreal, but sometimes forages on ground. Feeds on insects, reptiles, small mammals, carrion, and especially nestling birds. Eggs deposited in hole dug in ground, rotten stump, or termite mound.

Varanus yemenensis Böhme, Joger, Schatti 1989
Yemen Monitor

> 1989 *Varanus yemenensis* Böhme et al., Fauna Saudi Arabia 10:433

Description: Similar to *V. albigularis*, but with smaller scales, lack of pale spots or ocelli on the back. It has a conspicuous broad yellow band across the snout; tip of tail yellow. Total length: to 900 mm. (Plate 7.55)

Distribution: southwestern Arabian peninsula. (Map 7.26)

Ecology: Favors the bajadas of intermittent streams (wadis). Feeds on insects (primarily beetles), snails, and other invertebrates. Probably feeds on small vertebrates occasionally.

RANGE MAPS

Note: These maps provide general locations only and are not noted for specific areas. Range limits are approximate on these maps; distribution is not uniform within the indicated area.

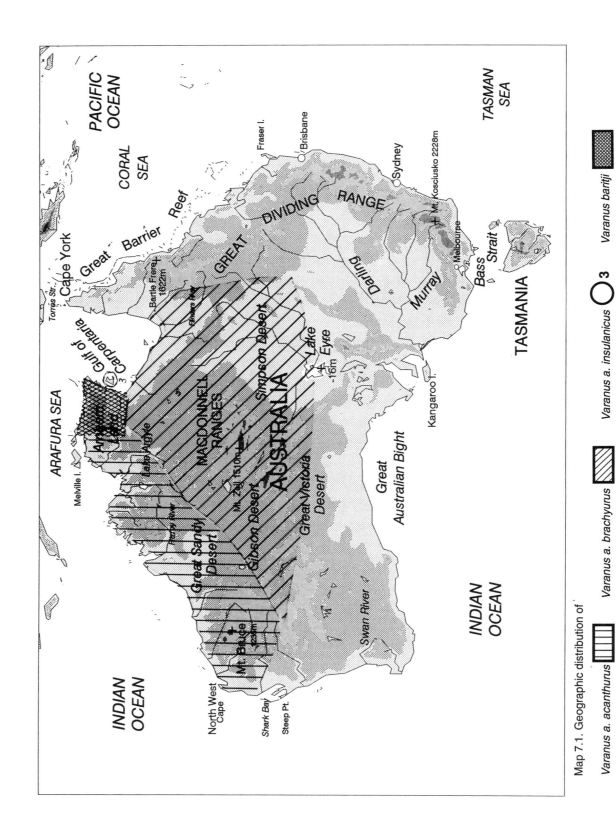

Map 7.1. Geographic distribution of *Varanus a. acanthurus*, *Varanus a. brachyurus*, *Varanus a. insulanicus*, *Varanus baritji*

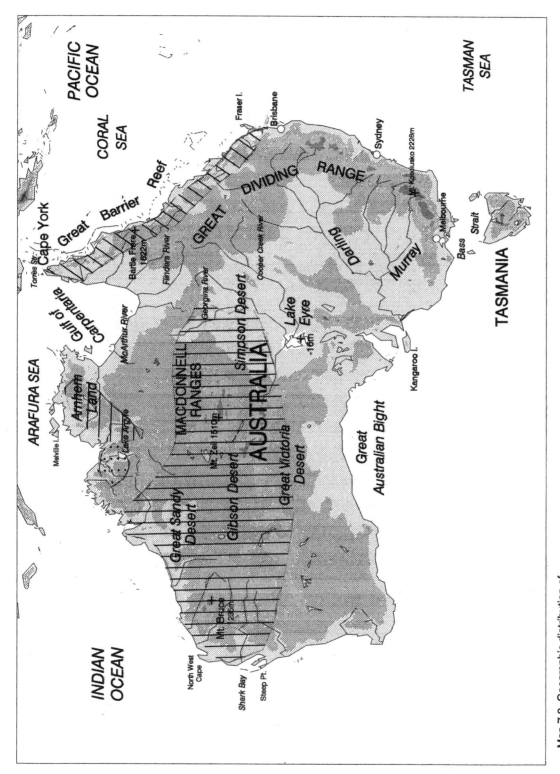

Map 7.2. Geographic distribution of *Varanus brevicauda*, *Varanus kingorum*, *Varanus primordius*, *Varanus semiremex*

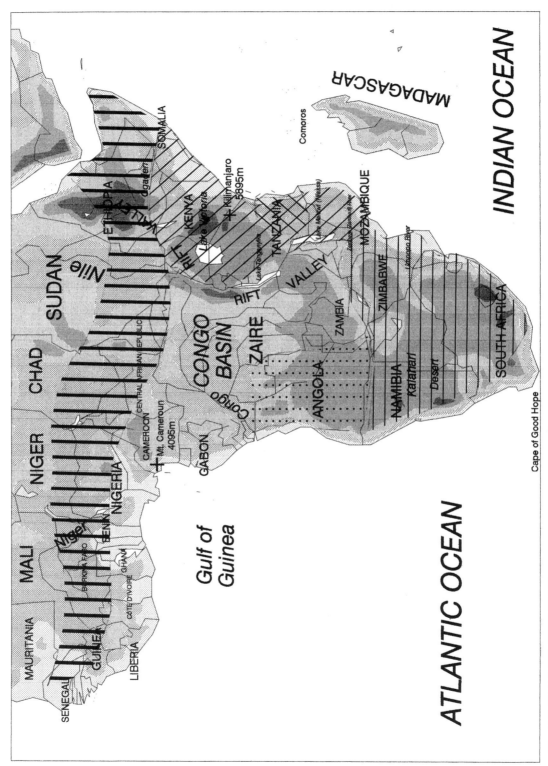

Map 7.3. Geographic distribution of *Varanus a. albigularis*, *Varanus a. angolensis*, *Varanus a. ionides*, *Varanus a. microstictus*, *Varanus exanthematicus*

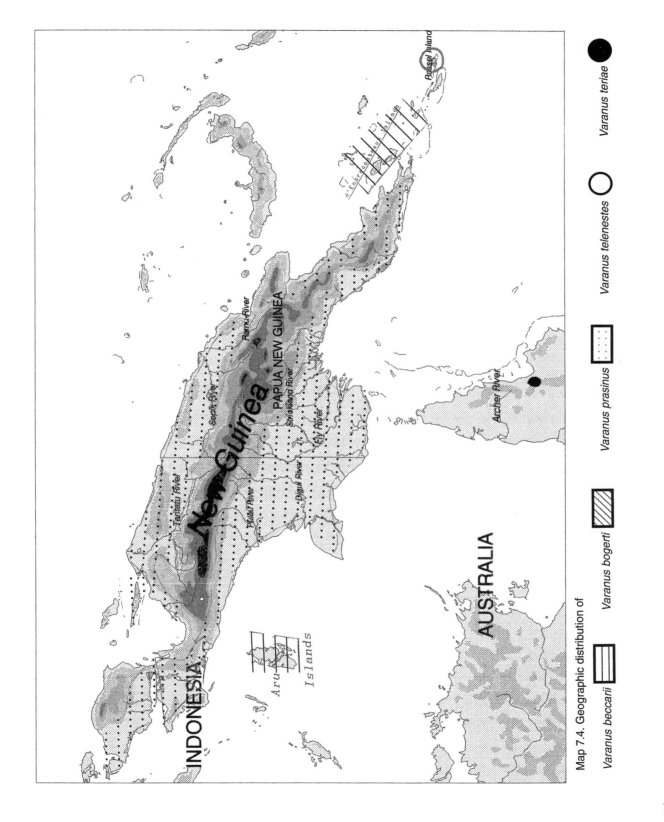

Map 7.4. Geographic distribution of *Varanus beccarii*, *Varanus bogerti*, *Varanus prasinus*, *Varanus telenestes*, *Varanus teriae*

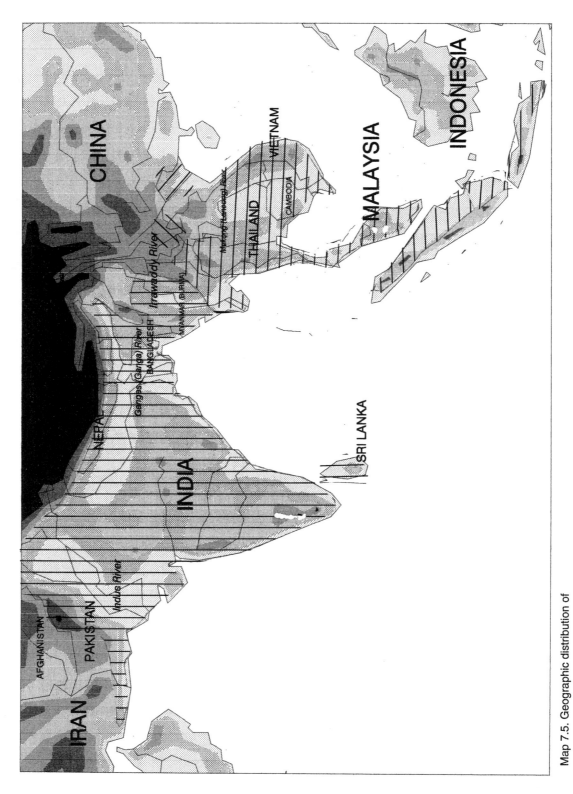

Map 7.5. Geographic distribution of *Varanus b. bengalensis*, *Varanus b. irrawadicus*, *Varanus b. nebulosus*, *Varanus b. vietnamensis*

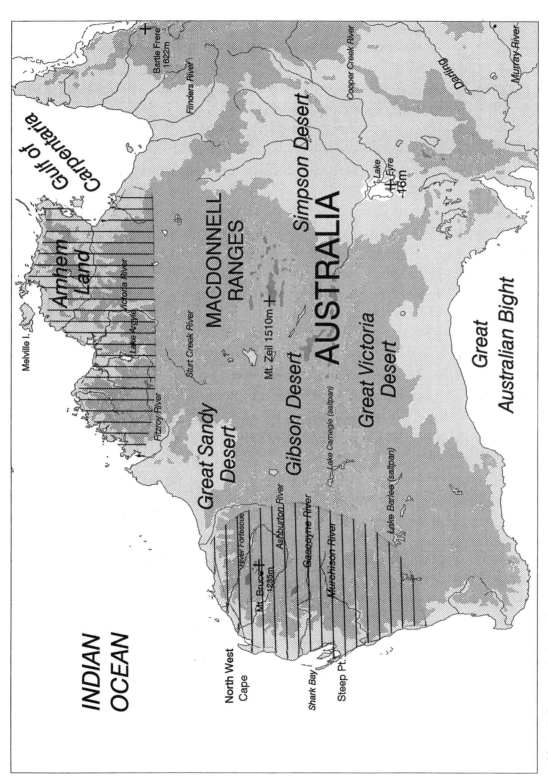

Map 7.6. Geographic distribution of *Varanus caudolineatus* and *Varanus mitchelli*

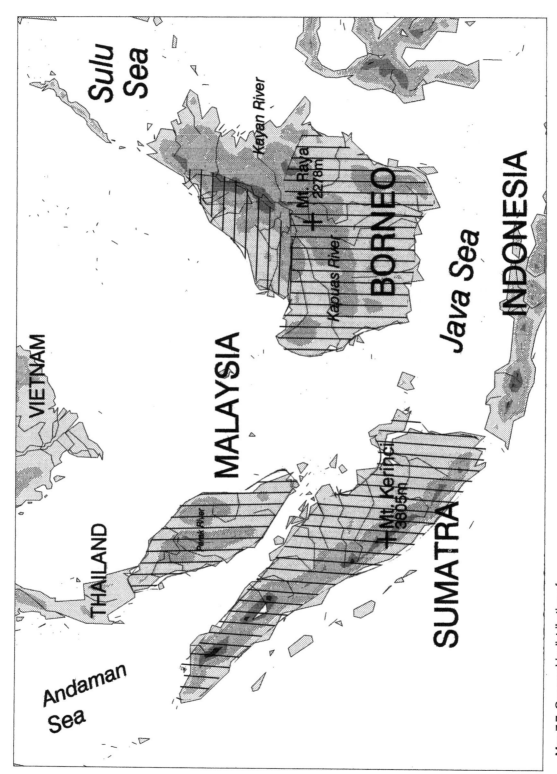

Map 7.7. Geographic distribution of *Varanus d. dumerilii* and *Varanus d. heteropholis*

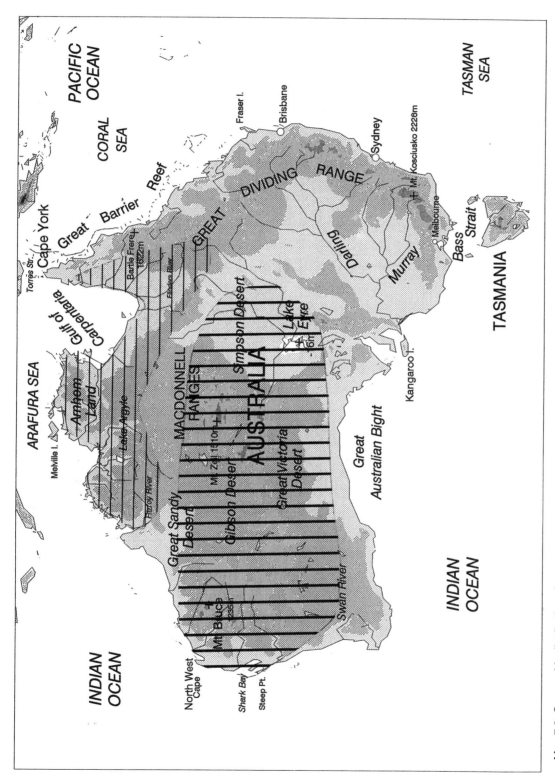

Map 7.8. Geographic distribution of *Varanus eremius*, *Varanus mertensi*

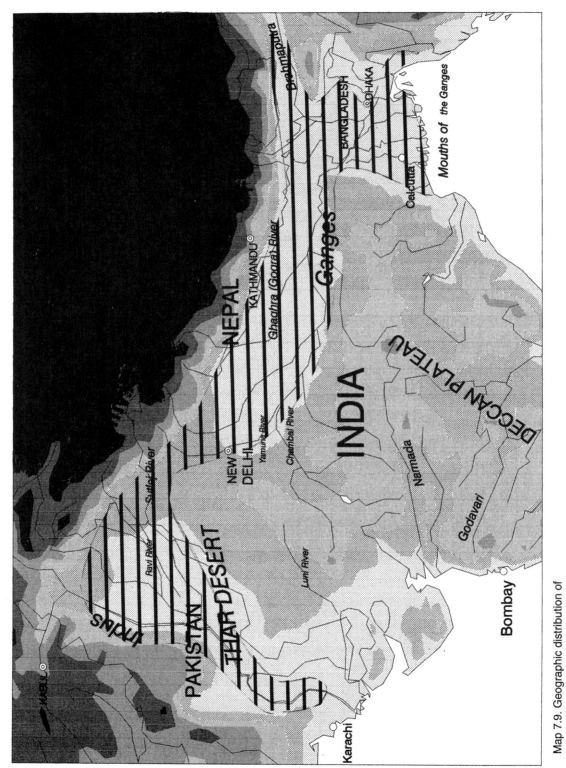

Map 7.9. Geographic distribution of *Varanus flavescens*

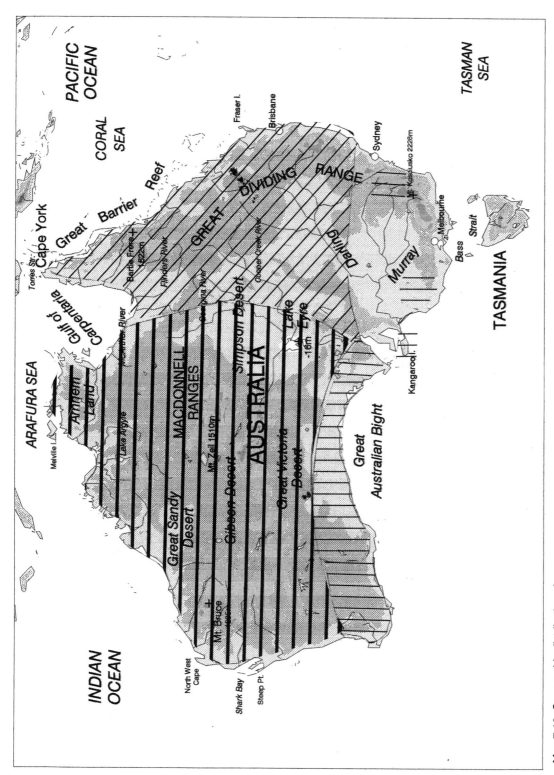

Map 7.10. Geographic distribution of *Varanus f. flavirufus*, *Varanus f. gouldii*, *Varanus rosenbergi*

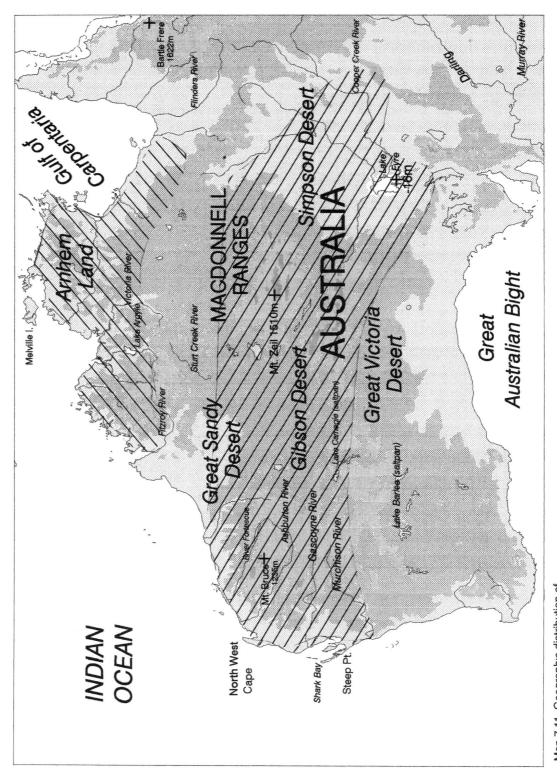

Map 7.11. Geographic distribution of *Varanus giganteus* and *Varanus glebopalma*

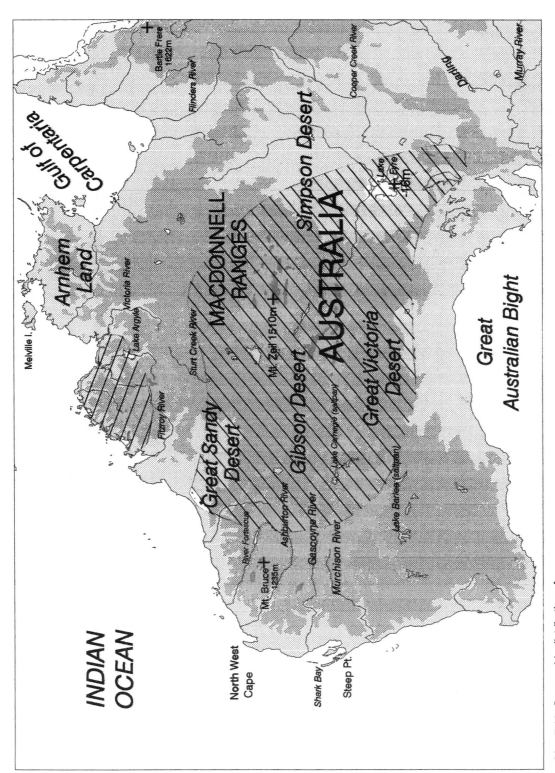

Map 7.12. Geographic distribution of *Varanus gilleni* and *Varanus glauerti*

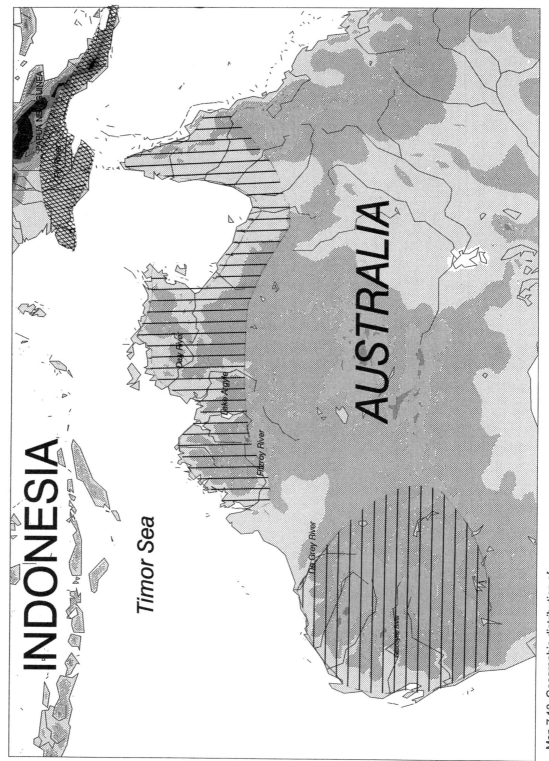

Map 7.13. Geographic distribution of *Varanus g. gouldii*, *Varanus g. horni*, *Varanus g. rubidus*

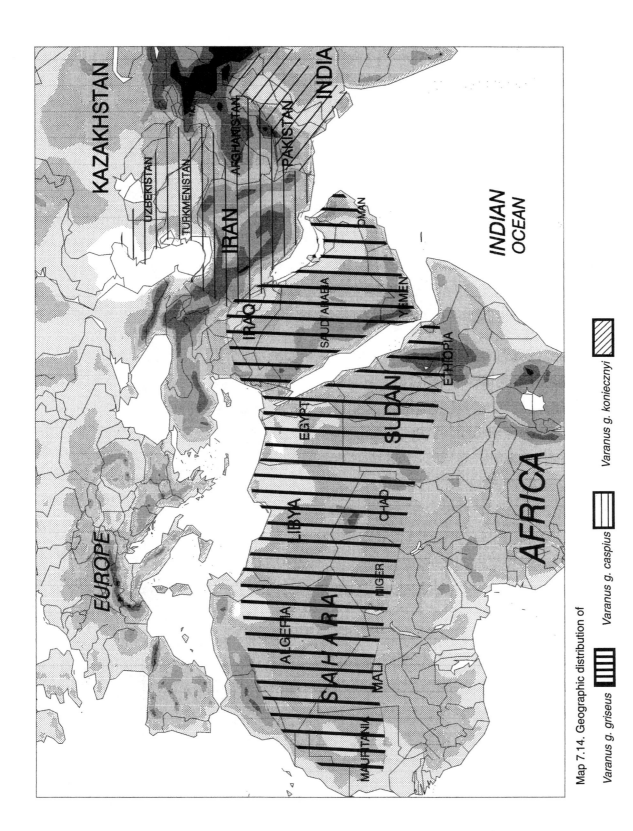

Map 7.14. Geographic distribution of *Varanus g. griseus*, *Varanus g. caspius*, *Varanus g. koniecznyi*

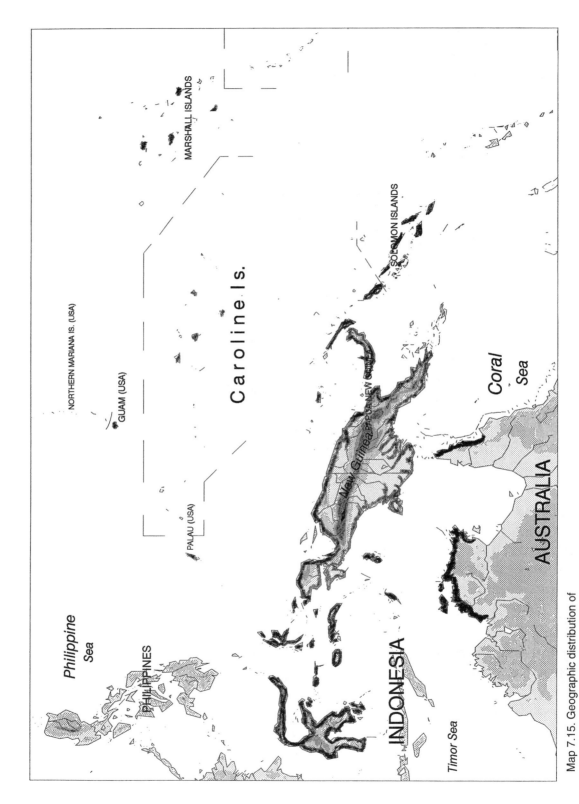

Map 7.15. Geographic distribution of *Varanus indicus* ■ (indicates coastlines)

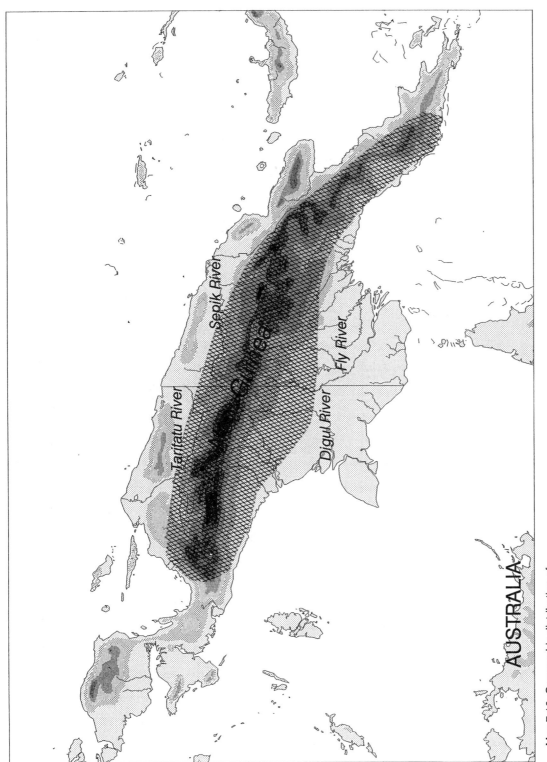

Map 7.16. Geographic distribution of *Varanus salvadorii*

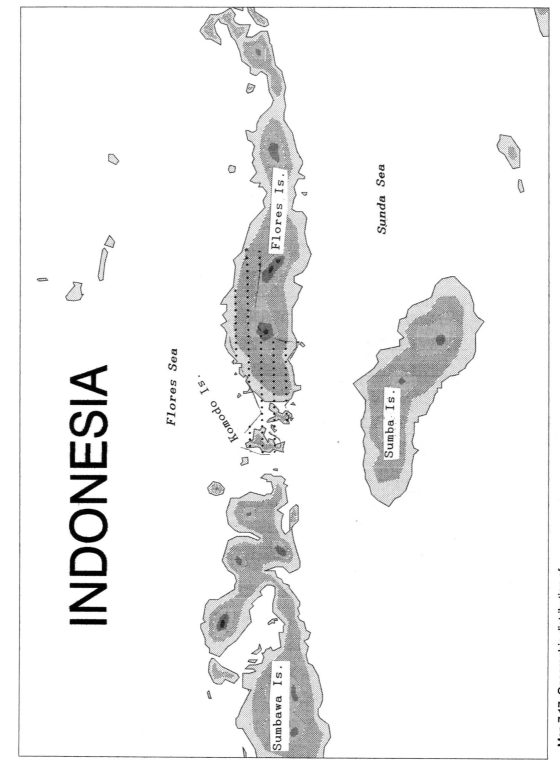

Map 7.17. Geographic distribution of *Varanus komodoensis*

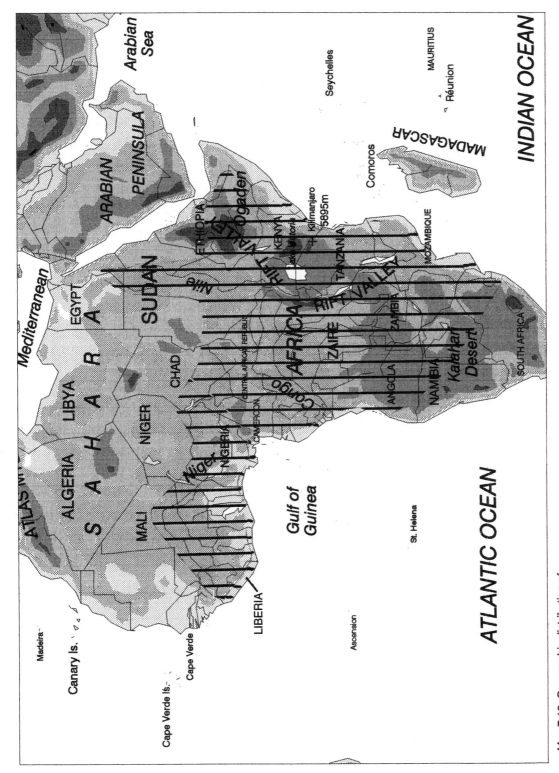

Map 7.18. Geographic distribution of
Varanus niloticus

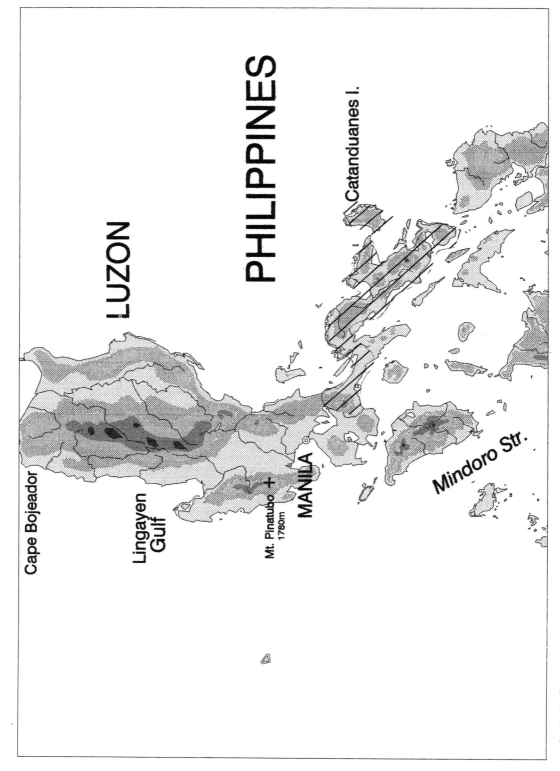

Map 7.19. Geographic distribution of *Varanus olivaceus*

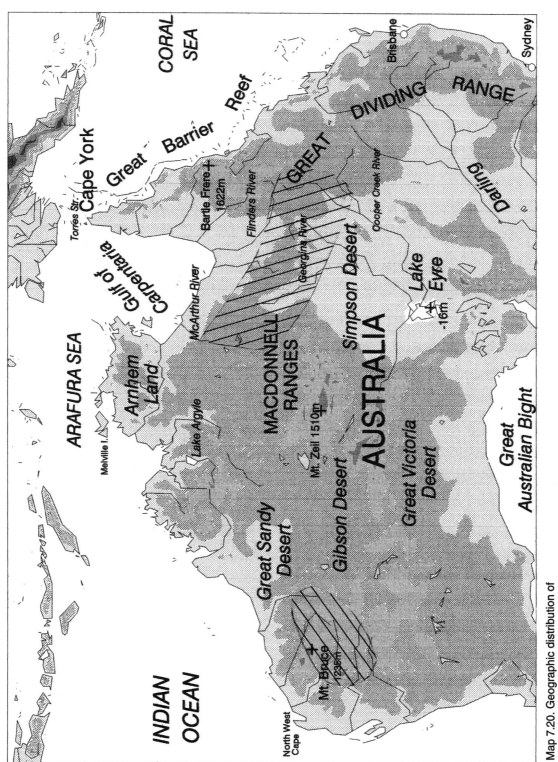

Map 7.20. Geographic distribution of *Varanus pilbarensis* *Varanus spenceri*

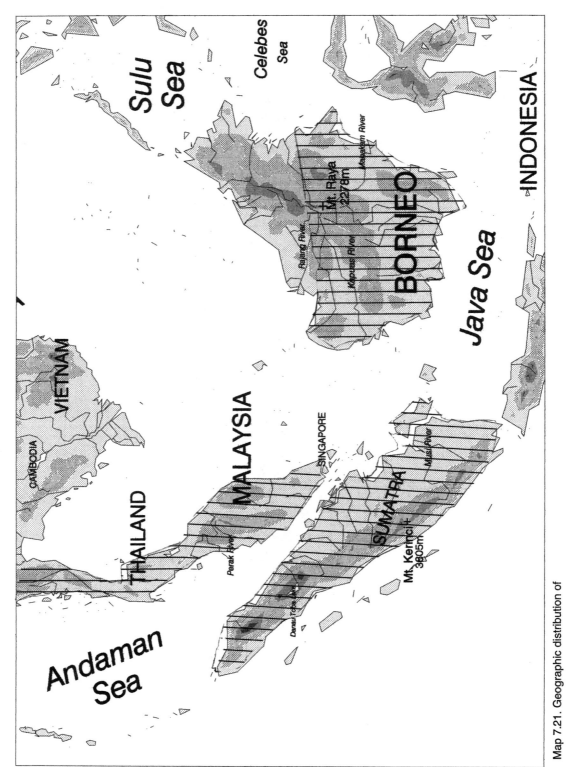

Map 7.21. Geographic distribution of
Varanus rudicollis

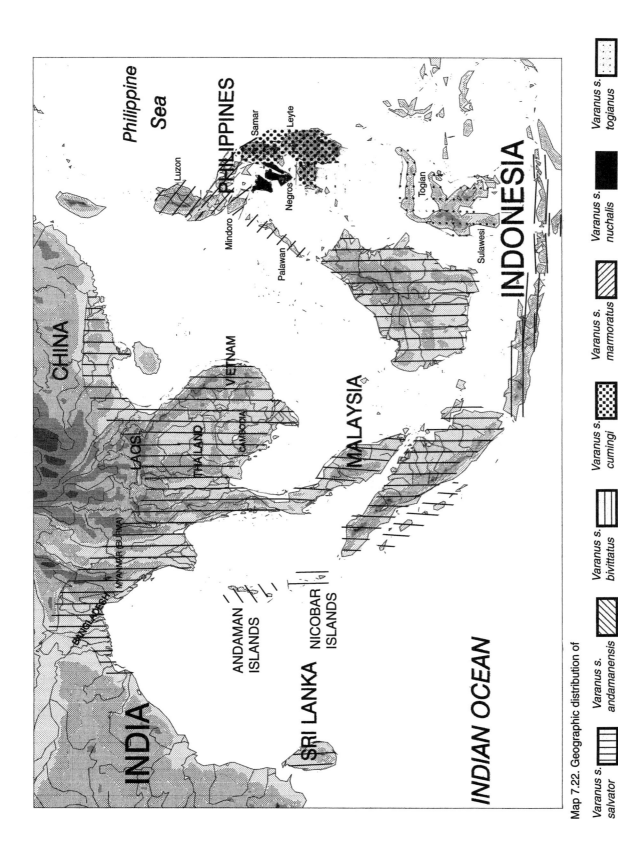

Map 7.22. Geographic distribution of *Varanus s. salvator*, *Varanus s. andamanensis*, *Varanus s. bivittatus*, *Varanus s. cumingi*, *Varanus s. marmoratus*, *Varanus s. nuchalis*, *Varanus s. togianus*.

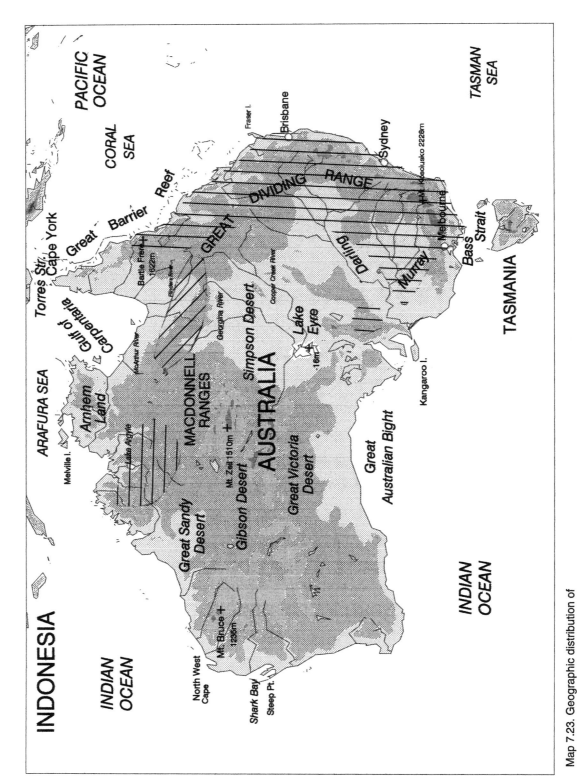

Map 7.23. Geographic distribution of *Varanus s. storri*, *Varanus s. ocreatus*, *Varanus varius*.

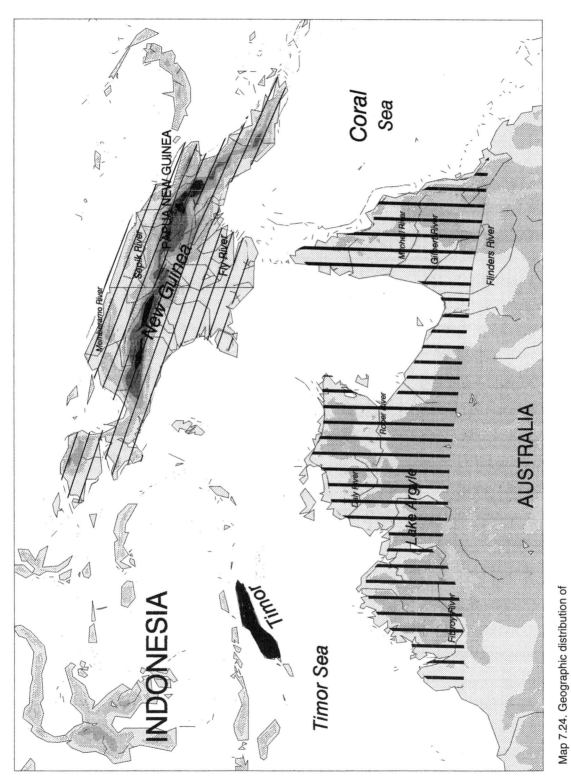

Map 7.24. Geographic distribution of *Varanus scalaris*, *Varanus t. timorensis*, *Varanus t. similis*.

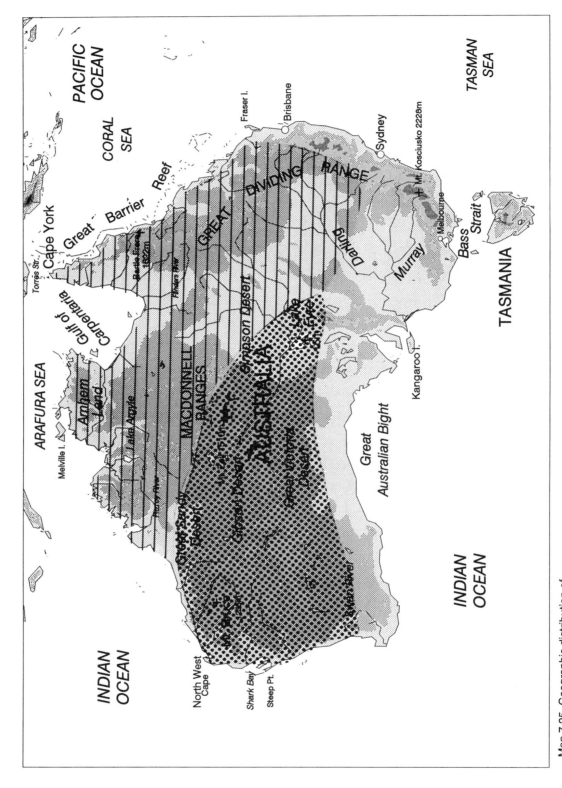

Map 7.25. Geographic distribution of *Varanus t. tristis* and *Varanus t. orientalis*

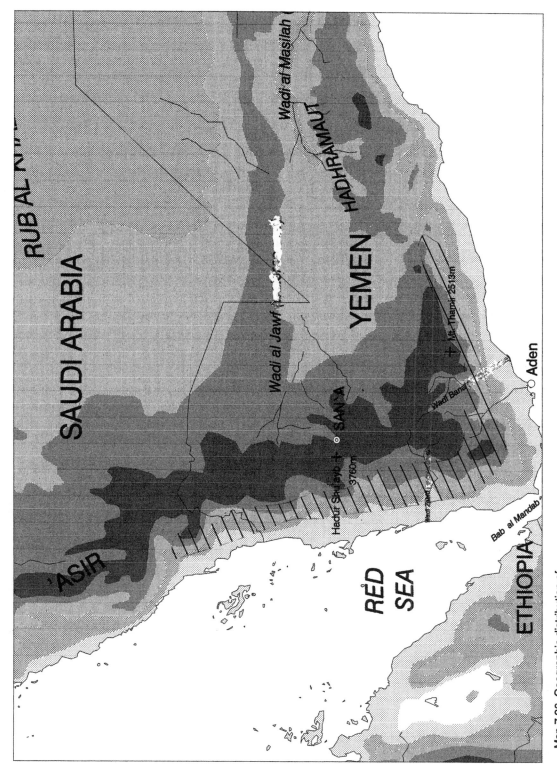

Map 7.26. Geographic distribution of *Varanus yemenensis*

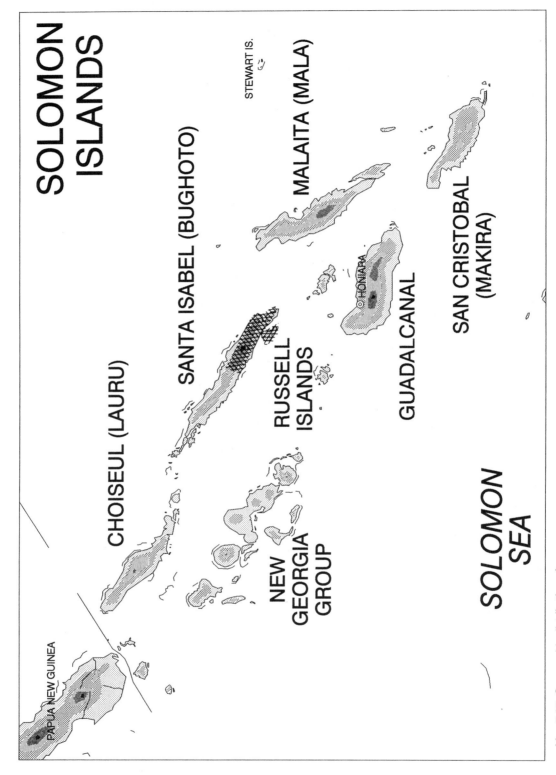

Map 7.27. Geographic distribution of *Varanus spinulosus*

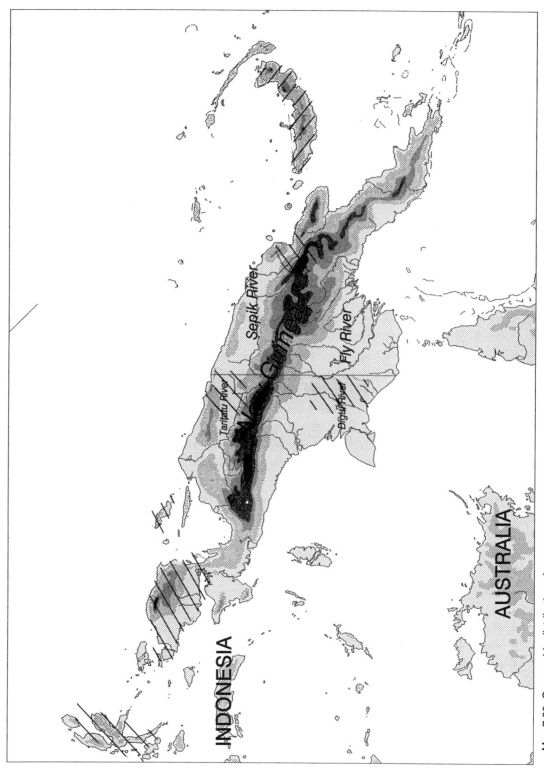

Map 7.28. Geographic distribution of *Varanus d. doreanus* and *Varanus d. finschi*

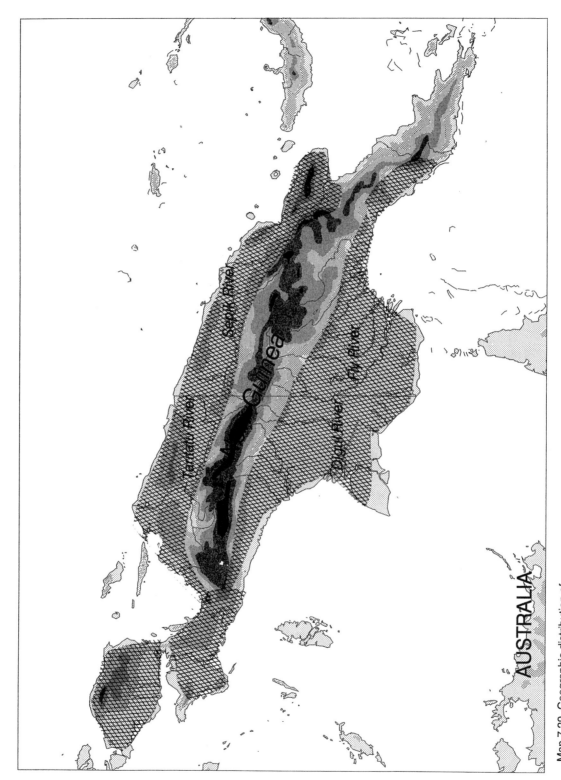

Map 7.29. Geographic distribution of *Varanus jobiensis*

Chapter 8
THE FUTURE FOR VARANIDS

A cursory reading of the earlier chapters of this book will indicate how little is yet known of the natural history of all but a handful of species of *Varanus*. This chapter is meant to be a summary of the needs for further research into varanid biology.

KNOWLEDGE GAPS

The study of good natural history must begin with an adequate taxonomy. In studying a group of organisms, it is vital to know where one species begins and another ends. The taxonomy of varanids has undergone numerous changes at the specific level in the last decade. There remains much more to be learned about the systematics of these animals whose classical morphology is so similar as to have more than once misled the eminent Robert Mertens, when all he had to guide him were skeletons and preserved specimens. Most of the advances in varanid systematics will come from people working with sympatric and insular populations who can do comparative analysis of the living lizards.

We already know of one new species from the Sir Edward Pelew Islands in the Gulf of Carpenteria and from adjacent mainland Australia. This will probably be named *V. "pelewensis,"* although it has not yet been formally described (D. King, pers. comm.). There are probably additional races of the *prasinus* group undescribed. One type from the Merauke River region of southern New Guinea may be enough different from the nominal race to warrant at least subspecific status (F. Yuwono, pers. comm.).

Similarly there are several reported "different looking" populations among the *indicus* and *salvator* groups in the islands of southeast Asia.

As the benefits of molecular biology techniques spread into herpetology systematics, DNA analysis is becoming more and more important. Currently it is the cytochrome-b gene of mitochondrial DNA (mDNA) that is most commonly used for evolutionary partitioning at the species/subspecies level. To date there have been no published studies using this technique on varanids, but it is reported that such studies are under way in at least one lab in the U.S.

A major evolutionary question that remains unresolved is the relationship between monitors and snakes. Are the similarities results of convergent evolution (the most popular hypothesis today) or are they a result of close common ancestry, as has been held by many over the years?

In the realm of lizard physiology, there are a number of projects to be undertaken. The study of water and electrolyte homeostasis has just begun. Comparative studies of species across habitat types need to be initiated. What physiological adaptations have appeared in desert species as compared with tropical rain forest species? None of the more arboreal species has yet been studied in this or most any aspect of its physiology.

Only a few species have been tested on a treadmill, mostly to determine aerobic scope. What is the comparative running speeds of these morphologically similar lizards? How does original habitat type influence running speed? What can these tests reveal about the animals' molecular energetics?

Metabolic rate studies have been done on 10 species. Such studies should be extended to other species, especially aquatic and arboreal forms. Such comparative work may yield real insights into the evolution of these fascinating lizards.

Ecological field studies need to be extended to more species. The use of radiotelemetry will undoubtedly answer many of the questions about activity patterns, core activity areas, predation, and other natural history aspects of lesser known species. Advances in the electronics of radiotelemetry will undoubtedly allow us to ask questions we do not even think of now. These studies would go a long way to expanding the tables of comparative data which today are often quite brief.

Dietary analysis, by stomach pumping, of live animals in the wild is needed to compare with studies from museum specimens. Seasonal studies may well broaden even further the range of foods taken by wild varanids. How accurate really is the portrayal of varanids as wide-ranging foragers?

Behavioral observation studies are in their infancy. Courtship and mating have been fully observed in only seven species. How universal are the courtship rituals? Do arboreal species court and/or mate on the ground or in trees? What modifications in the stereotypical ritual have been required? How do the various species sort out in the three types of mating systems so far described? Do females return to the location of their own hatching to lay their own eggs several years later, thus explaining some of the clustering of nests that has been observed?

Do females have particular basking sites or foraging paths that keep them close to the nest during incubation? The circumstantial evidence of the mother, in some species at least, returning to the nest many months after she deposited the eggs, to free the hatchlings is just beginning to appear. How widespread is this behavior in the genus? Is it used only by individuals nesting in termitaria?

Since hatchlings of most species are rarely seen in the wild, where *do* very young monitors spend most of their time?

Growth rate data in the wild is almost totally lacking. It is to be hoped that some of the long-term studies can be turned over to local naturalists to obtain this much needed data set. Such studies could be expanded to yield other as of now nonexistent information on survivorship and longevity in wild populations. No studies as yet have reached the point of tracing individual lizards throughout their life.

Ecological studies for conservation purposes especially need to focus on two important questions. How large a core activity area do varanids of various species require? What minimal kinds of natural vegetation must be included in sustainable habitat? Population density data are urgently needed if any kind of sustained-yield populations are to remain in many locales as a renewable resource for the human inhabitants.

The question of sex ratios is another population parameter just beginning to be explored.

The husbandry literature continually mentions the heavy parasitemia of monitors. No scientifically valid studies, however, have been done to determine whether the supposed heavy internal parasite load is real or apparent. Veterinarians have been slow to determine and publish treatment regimes.

Captive propagation may in the long run be one of the most important projects for the future of varanids. The demand for skins is not likely to slow any time soon; and, as we have seen, legislation in this regard is ineffective. Ranching for skins has succeeded in easing the pressure on wild populations of those other reptiles whose skins are in much demand by the leather industry —the crocodilians. There are still elementary unanswered questions about the preprequisites for successful captive breeding of varanids. It will undoubtedly be the combined efforts of private breeders and zoos that will find the answers to those questions. There are indications that once the breeding problems have been solved, female varanids can be induced to lay several clutches of eggs each year.

CONTRIBUTIONS FROM CAPTIVE STUDIES

Captive varanids offer an opportunity to study aspects of their behavior which are just not observable in the wild for most species. Auffenberg (1988) used captive *V. olivaceus* to obtain much of the data on behavior for his monograph on that species, and his 1994 work on *V. bengalensis* relies heavily on observations made on a captive group he kept at his Florida home. Such phenomena as sleeping and waking periods, walking and climbing speeds, and behavior related to feeding, courtship and combat all lend themselves to analysis in captive specimens. One great advantage of studying such behaviors in captive animals is that comparative data can be gathered under controlled conditions.

Analysis of these comparative ethologies should help systematists in sorting out the species puzzles of so many lizards with similar morphologies. Some of the physiological questions still unanswered for most species can be addressed in a laboratory setting before expensive field studies are initiated.

Finally, it will be among herpetoculturalists that the puzzles and problems of captive breeding are solved. Progress is already being made (Ratnam, 1994). And therein may be the salvation of those species being decimated by the skin trade.

APPENDIX A

Recognized Taxa of the Genus
Varanus
Cross-Referenced by Scientific Name

V. acanthurus acanthurus	Northwestern Ridge-tailed monitor
V. acanthurus brachyurus	Common Ridge-tailed monitor
V. acanthurus insulanicus	Island Ridge-tailed Monitor
V. albigularis albigularis	South African White-throated Monitor
V. albigularis angolensis	Angola White-throated Monitor
V. albigularis ionidesi	Ionides' White-throated Monitor
V. albigularis microstictus	East African White-throated Monitor
V. baritji	White's Monitor
V. beccarii	Black Tree Monitor
V. bengalensis bengalensis	Bengal Monitor
V. bengalensis irrawadicus	Yunnan Monitor
V. bengalensis nebulosus	Clouded Monitor
V. bengalensis vietnamensis	Vietnam Monitor
V. bogerti	Bogert's Tree Monitor
V. brevicauda	Short-tailed Monitor
V. caudolineatus	Stripe-tailed Monitor
V. doreanus doreanus	Blue-tailed Monitor
V. doreanus finschi	Finsch's Monitor
V. dumerilii dumerilii	Dumeril's Monitor
V. dumerilii heteropholis	Sarawak Forest Monitor
V. eremius	Rusty Desert Monitor
V. exanthematicus	Savannah Monitor
V. flavescens	Yellow Monitor
V. flavirufus flavirufus	Sand Monitor
V. flavirufus gouldii	Bungarra
V. giganteus	Perentie
V. gilleni	Pygmy Mulga Monitor
V. glauerti	Glauert's Monitor
V. glebopalma	Long-tailed Rock Monitor
V. gouldii gouldii	Gould's Monitor
V. gouldii horni	Horn's Monitor
V. gouldii rubidus	Yellow-spotted Monitor
V. griseus griseus	Western Desert Monitor
V. griseus caspius	Eastern Desert Monitor
V. griseus koniecznyi	Thar Desert Monitor
V. indicus	Mangrove Monitor

V. jobiensis	Peach-throated Monitor
V. kingorum	Kings' Monitor
V. komodoensis	Komodo Monitor
V. mertensi	Mertens' Water Monitor
V. mitchelli	Mitchell's Water Monitor
V. niloticus	Nile Monitor
V. olivaceus	Gray's Monitor
V. pilbarensis	Pilbara Monitor
V. prasinus	Green Tree Monitor
V. primordius	Northern Blunt-spined Monitor
V. rosenbergi	Rosenberg's Monitor
V. rudicollis	Rough-necked Monitor
V. salvadorii	Papua Monitor
V. salvator salvator	Asian Water Monitor
V. salvator andamanensis	Andaman Islands Water Monitor
V. salvator bivittatus	Two-striped Water Monitor
V. salvator cumingi	Cuming's Water Monitor
V. salvator marmoratus	Marbled Water Monitor
V. salvator nuchalis	Negros Water Monitor
V. salvator togianus	Togian Water Monitor
V. scalaris	Australian Spotted Tree Monitor
V. semiremex	Rusty Monitor
V. spenceri	Spencer's Monitor
V. spinulosus	Solomons Keeled Monitor
V. storri storri	Eastern Storr's Monitor
V. storri ocreatus	Western Storr's Monitor
V. telenestes	Rossel Island Monitor
V. teriae	Cape York Tree Monitor
V. timorensis timorensis	Timor Monitor
V. timorensis similis	New Guinea Spotted Tree Monitor
V. tristis tristis	Black-headed Monitor
V. tristis orientalis	Freckled Monitor
V. varius	Lace Monitor
V. yemenensis	Yemen Monitor

APPENDIX B

Recognized Taxa of the Genus
Varanus
Cross-Referenced by Common Name

Bengal Monitor	*V. bengalensis bengalensis*
Black-headed Monitor	*V. tristis tristis*
Blue-tailed Monitor	*V. doreanus doreanus*
Bungarra	*V. flavirufus gouldii*
Clouded Monitor	*V. bengalensis nebulosus*
Desert Monitor, Eastern	*V. griseus caspius*
Desert Monitor, Thar	*V. griseus koniecznyi*
Desert Monitor, Western	*V. griseus griseus*
Dumeril's Monitor	*V. dumerilii dumerilii*
Finsch's Monitor	*V. doreanus finschi*
Freckled Monitor	*V. tristis orientalis*
Glauert's Monitor	*V. glauerti*
Gould's Monitor	*V. gouldii gouldii*
Gray's Monitor	*V. olivaceus*
Horn's Monitor	*V. gouldii horni*
Kings' Monitor	*V. kingorum*
Komodo Monitor	*V. komodoensis*
Lace Monitor	*V. varius*
Long-tailed Rock Monitor	*V. glebopalma*
Mangrove Monitor	*V. indicus*
Nile Monitor	*V. niloticus*
Northern Blunt-spined Monitor	*V. primordius*
Papua Monitor	*V. salvadorii*
Peach-throated Monitor	*V. jobiensis*
Perentie	*V. giganteus*
Pilbara Monitor	*V. pilbarensis*
Pygmy Mulga Monitor	*V. gilleni*
Ridge-tailed Monitor, Common	*V. acanthurus brachyurus*
Ridge-tailed Monitor, Island	*V. acanthurus insulanicus*
Ridge-tailed Monitor, Northwestern	*V. acanthurus acanthurus*
Rosenberg's Monitor	*V. rosenbergi*
Rossel Island Monitor	*V. telenestes*
Rough-necked Monitor	*V. rudicollis*
Rusty Desert Monitor	*V. eremius*
Rusty Monitor	*V. semiremex*
Sand Monitor	*V. flavirufus flavirufus*

Sarawak Forest Monitor	*V. dumerilii heteropholis*
Savannah Monitor	*V. exanthematicus*
Short-tailed Monitor	*V. brevicauda*
Solomons Keeled Monitor	*V. spinulosus*
Spencer's Monitor	*V. spenceri*
Storr's Monitor, Eastern	*V. storri storri*
Storr's Monitor, Western	*V. storri ocreatus*
Stripe-tailed Monitor	*V. caudolineatus*
Timor Monitor	*V. timorensis timorensis*
Tree Monitor, Australian Spotted	*V. scalaris*
Tree Monitor, Black	*V. beccarii*
Tree Monitor, Bogert's	*V. bogerti*
Tree Monitor, Cape York	*V. teriae*
Tree Monitor, Green	*V. prasinus*
Tree Monitor, New Guinea Spotted	*V. timorensis similis*
Vietnam Monitor	*V. bengalensis vietnamensis*
Water Monitor, Andaman Islands	*V. salvator andamanensis*
Water Monitor, Asian	*V. salvator salvator*
Water Monitor, Cuming's	*V. salvator cumingi*
Water Monitor, Marbled	*V. salvator marmoratus*
Water Monitor, Mertens'	*V. mertensi*
Water Monitor, Mitchell's	*V. mitchelli*
Water Monitor, Negros	*V. salvator nuchalis*
Water Monitor, Togian	*V. salvator togianus*
Water Monitor, Two-striped	*V. salvator bivittatus*
White's Monitor	*V. baritji*
White throated Monitor, Angola	*V. albigularis angolensis*
White-throated Monitor, East African	*V. albigularis microstictus*
White-throated Monitor, Ionides'	*V. albigularis ionidesi*
White-throated Monitor, South African	*V. albigularis albigularis*
Yellow Monitor	*V. flavescens*
Yellow-spotted Monitor	*V. gouldii rubidus*
Yemen Monitor	*V. yemenensis*
Yunnan Monitor	*V. bengalensis irrawadicus*

REFERENCES

Adler, K. 1989. Herpetologists of the past. *In* K. Adler (ed.) Contributions to the History of Herpetology, No. 5. SSAR publ.

Allen, M.E., and O.T. Ofledal. 1994. The nutrition of carnivorous reptiles, pp. 71–82 *In* J.R. Murphy, K. Adler, J.T. Collins (eds.), Captive Management and Conservation of Amphibians and Reptiles. SSAR, Contr. Herpet. 11.

Alpin, K.P., and H.R. Harding. 1994 Ultrastructural observations on spermatozoa of four species of Australian varanids, p. 8, *In* Abstr. 2nd World Congress Herpet., Adelaide.

Andrews, H.V., and M. Gaulke. 1990. Observations on the reproductive biology and growth of the water monitor (*Varanus salvator*) at the Madras Crocodile Bank. Hamadryad 15:1–5.

Annandale, T. 1903. *In* G.A. Boulenger, Report on the batrachians and reptiles. Fasc. Malay. Zool. 1:131–176.

Anon. 1992. The complete guide to keeping monitors. Reptile News Press, 41 pp.

Auffenberg, W. 1981a. The behavioral ecology of the Komodo monitor. Univ. of Florida Press, Gainesville, 406 pp.

———. 1981b. Combat behaviour in *Varanus bengalensis* (Sauria: Varanidae). J. Bombay Nat. Hist. Soc. 7:54–72.

———. 1983. Courtship behavior in *Varanus bengalensis* (Sauria: Varanidae). *In* A. G. J. Rhodin and K. Miyata (eds.). Advances in Herpetology and Evolutionary Biology: Essays in Honor of Ernest E. Williams. Cambridge, Mus. Comp. Zool., Harvard U.

———. 1984. Notes on the feeding behaviour of *Varanus bengalensis* (Sauria: Varanidae). J. Bombay Nat. Hist. Soc. 80:286–302.

———. 1988. Gray's monitor lizard. Univ. Florida Press, Gainesville, 419 pp.

———. 1994. The Bengal monitor. Univ. Florida Pres, Gainesville, 592 pp.

———, and I. M. Ipe. 1983. The food and feeding of juvenile bengal monitor lizards (*Varanus bengalensis*). J. Bombay Nat. Hist. Soc. 80:111–124.

———, H. Rehman, F. Iffat, Z. Perveen. 1989. A study of *Varanus flavescens* (Hardwicke & Gray) (Sauria: Varanidae). J. Bombay Nat. Hist. Soc. 86:286–307.

———, ———, ———, ———. 1990. Notes on the biology of *Varanus griseus koniecznyi* Mertens (Sauria: Varanidae). J. Bombay Nat. Hist. Soc. 87:26–36.

———, Q. N. Arain, N. Khurshid. 1991. Preferred habitat, home range and movement patterns of *Varanus bengalensis* in southern Pakistan. Mertensiella 2:7–28.

Baird, I. L. 1970. The anatomy of the reptilian ear. *In* C. Gans and T.S. Parsons (eds.), Biology of the Reptilia, vol.2, pp. 193–275.

Balsai, M. 1992. The general care and maintenance of savannah monitors and other popular monitor species. Advanced Aquarium Systems, Lakeside, CA, 55 pp.

Barker, D.G. 1985. Maintenance and reproduction of green tree monitors at the Dallas Zoo. pp. 91–92 In Proc. 8th Ann. Reptile Symp. Captive Propagation and Husbandry.

Barnett, B. 1979. Incubation of sand goanna (*Varanus gouldii*) eggs. Herpetofauna 11:21–22.

Bartholomew, G. A., and V .A. Tucker. 1964. Size, body temperature, thermal conductance, oxygen consumption, and heart rate in Australian varanid lizards. Physiol. Zool. 37:341–354.

Bartlett, R. D. 1982. Initial observations on the captive reproduction of *Varanus storri* Mertens. Herpetofauna 13:6–7.

Baverstock, P. R., D. King, M. King, J. Birrell, M. Krieg. 1993. The evolution of the Varanidae: microcomplement fixation analysis of serum albumins. Aust. J. Zool. 41:621–638.

Becker, H. O., W. Bohme, S. F. Perry. 1989. Die lungenmorphologie der warane (Reptilia:Varanidae) und ihre systematisch-stammesgeschichtliche bedeutung. Bonn. zool. butr. 40:27–56.

Behrmann, H. J. 1981. Haltung und nachzucht von *Varanus t. timorensis* (Reptilia:Sauria:Varanidae). Salamandra 17:198–201.

Bels, V .L., J. P. Gasc, S. Renous, R. Vernet. 1994. Throat display in varanids: eco-ethological and functional studies, p. 23 In Abstr. 2nd World Congress Herpet., Adelaide.

Bickler, P. E., and R. A. Anderson. 1986. Ventilation, gas exchange, and aerobic scope in a small monitor lizard, *Varanus gilleni*. Physiol. Zool. 59:76–83.

Böhme, W. 1988a. Zur genitalmorphologie der Sauria. Bonn. zool. Monogr. 27:1–176.

———, W. 1988b. The Argus monitor (*Varanus panoptes*, Storr 1980) of New Guinea: *V. panoptes horni* ssp. n. Salamandra 24:87–101.

——— 1991a. The identity of *Varanus gouldii* (Gray, 1838), and the nomenclature of the *V. gouldii* species complex. Mertensiella 2:38–41.

——— 1991b. New findings on the hemipenial morphology of monitor lizards and their systematic implications. Mertensiella 2:42–49.

———, U. Joger, B. Schatti. 1989. A new monitor lizard (Reptilia:Varanidae) from Yemen, with notes on ecology, phylogeny and zoology. Fauna Saudi Arabia 10:433–448.

———, H. G. Horn, T. Ziegler. 1994. Zur taxonomie der Pazifikwarane (*Varanus indicus* komplex): Revalidierung von *Varanus doreanus* (A.B. Meyer, 1874) mit beschreibung einer neuen unterart. Salamandra 15:119–142.

Boulenger, G. A. 1885. Catalogue of the lizards in the British Museum, vol. 2, London.

Boulenger, G.A. 1903. Report on the batrachians and reptiles. Fasc. Malay. Zool. 1:131–176.

Boylan, T. 1995. Field observations, captive breeding and growth rates of the lace monitor, *Varanus varius*. Herpetofauna 25:10–14.

Boyle, D. M., and W. E. Lamoreaux. 1984. Captive reproduction in the pygmy mulga monitor, *Varanus gilleni*, at the Dallas Zoo. pp 59–63 *In* 7th Ann. Reptile Symp. on Captive Prop. and Husbandry.

Branch, W .R. 1991. The Regenia registers of 'Gogga' Brown (1869–1909) "Memoranda on a species of monitor or varan." Mertensiella 2:57–110.

Braysher, M., and B. Green. 1970. Absorption of water and electrolytes from the cloaca of an Australian lizard, *Varanus gouldii* (Gray). Comp. Biochem. Physiol. 35:607–614.

Bredl, J., and H. G. Horn. 1987. Uber die nachzucht des australischer riesenwarans *Varanus giganteus* (Gray 1845) (Sauria:Varanidae). Salamandra 23:90–96.

———, and T. D. Schwaner. 1988. First record of captive propagation of the lace monitor. *Varanus varius* (Sauria:Varanidae). Herpetofauna 15:20–21.

Brotzler, A. 1965. Mertens-wasserwarane (*Varanus mertensi* Glauert 1951) zuchteten in der Wilhelma. Freunde Koln Zoo 8:89.

Burggren, W. W. 1987. Form and function in reptilian circulations. Amer. Zool. 27:5–19.

Busono, M. S. 1974. Facts about the *Varanus komodoensis* at the Gembiro Loka Zoo at Yogyakarta. Zool. Gart., Jena 44:62–63.

Card, W. 1994. Double clutching Gould's monitors (*Varanus gouldii*) and Gray's monitors (*Varanus olivaceus*) at the Dallas Zoo. Herp. Rev. 25:111–114.

———, and A.G. Kluge. 1995. Hemipeneal skeleton and varanid lizard systematics. J. Herpet. 29:275–280.

Carroll, R. L., and M. deBraga. 1992. Aigialosaurs: Mid-Cretaceous varanoid lizards. Jour. Vert. Paleontol. 12:66–86.

Carter, D. B. 1990. Courtship and mating in wild *Varanus varius* (Varanidae: Australia). Mem. Queensland Mus. 29:333–338.

———. 1994. A novel reproductive strategy by a monitor lizard, p. 49, *In* Abstr. 2nd World Congress Herpet., Adelaide.

Christian, K., B. Weavers, B. Green, L. K. Corbett. 1994. Seasonal activity and energetics of two species of tropical varanid lizards. Abstr. 2nd World Congress Herpet., Adelaide, :56

———, and B. Weavers. 1994. Analysis of the activity and energetics of the lizard *Varanus rosenbergi*. Copeia 1994:289–295.

Cisse, M. 1972. The diet of varanids in Senegal. Bull. l'Inst. Fond. d'Afrique Noire 34:503–515.

Cisse, M. 1976. The reproductive cycle of varanids in Senegal. Bull. Inst. Fond. Afr. Noire. Ser.A 38:188–205.

Cogger, H. G. 1992. Reptiles and amphibians of Australia. Cornell Univ. Press, Ithaca, N.Y., 6th ed, 775 pp.

Cott, H. B. 1961. Scientific results of an enquiry into the ecology and economic status of the Nile Crocodile (*Crocodylus niloticus*) in Uganda and Northern Rhodesia. Trans. Zool. Soc. London 29:211–356.

Cowles, R. B. 1930. The life history of *Varanus niloticus* (Linn.) as observed in Natal, South Africa. J. Entomol. Zool. 22:1–31.

Crebs, U. 1979. *Varanus dumerilii*—a specialized crab eater? Salamandra 15:146–157.

Daltry, J. 1991. The social hierarchy of the water monitor, *Varanus salvator*, in India and Sri Lanka. Hamadryad 16:10–20.

Daudin, F. M. 1802. Histoire naturelle generale et particuliere des reptiles, vol. 3, Paris.

———. 1803. Histoire naturelle generale et particuliere des reptiles, vol. 8, Paris.

David, R. 1970. Breeding the mugger crocodile and water monitor, *Crocodylus palustris* and *Varanus salvator*, at Ahmedabad Zoo. Int. Zoo Yb. 10:116–117.

Davis, R., R. Darling, A. Darlington. 1986. Ritualized combat in captive Dumeril's monitors, *Varanus dumerili*. Herp. Rev. 17:85–88.

de Buffrenil, V. 1994. Economic use and some biological parameters of heavily exploited populations of the Nile monitor (*Varanus niloticus niloticus*), p. 69, In Abstr. 2nd World Congress Herpet., Adelaide.

De Lisle, H. F. 1991. Behavioral ecology of the banded rock lizard (*Petrosaurus mearnsi*). Bull. So. Calif. Acad. Sci. 90:102–117.

Deranyagala, R. Y. 1958. Pseudocombat of the monitor *Varanus bengalensis*. Spol. Zeyl., Bull. Nat. Mus. Ceylon 28:11–13.

Diamond, J. 1992. The evolution of dragons. Discover 13(12):72–80.

Ditmars, R. 1910. Reptiles of the world. Macmillan, N.Y., 321 pp.

Djasmani and Rifanie. 1988. Survey of monitor lizards (*Varanus salvator*) in South Kalimantan. Unpubl. report of Univ. Lambung Mangkurat, Kelompok Program Studi.

Dryden, G. L., and E. D. Wikramanayake. 1991. Space and time sharing by *Varanus salvator* and *V. bengalensis* in Sri Lanka. Mertensiella 2:111–119.

Earll, C. R. 1982. Heating, cooling and oxygen consumption rates in *Varanus bengalensis*. Comp. Biochem. Physiol. 72A:377–381.

Edmund, A. G. 1969. Dentition. Pp. 117–200, In A. C. Gans, A. d'A. Bellairs, T. Parsons (eds.) Biology of the Reptilia, vol. 1, Morphology. Academic Press.

Ehmann, H. 1992. Encyclopedia of Australian Animals: Reptiles. Austr. Mus. Angus & Robertson publ., pp. 144–158

Eidenmuller, B. 1986. Observations on the care and a recent breeding of *Varanus (Odatria) timorensis timorensis* (Gray 1831). Salamandra 22:157–161.

———. 1990. Observations on keeping and breeding of *Varanus mertensi*. Salamandra 26:132–139.

———, and H.G. Horn. 1985. Some examples of breeding and the present state of knowledge about breeding in *Varanus (Odatria) storri* (Mertens 1966). Salamandra 21:55–61.

Erdelen, W. 1988. Survey of the status of the water monitor lizard (*Varanus salvator*; Reptilia:Varanidae) in south Sumatra. Unpubl. report.

Erdfelder, K. H. 1984. Haltung und zucht des stachelschwanwarans *Varanus acanthurus* Boulenger 1885. Sauria 6:9–11.

Estes, R., K. de Queiroz, J. Gauthier. 1988. Phylogenetic relationships within squamata. *In* R. Estes and G. Pregill (eds.) Phylogenetic Relationships of the Lizard Families, Stanford U. Press.

Frye, F. L. 1991. Biomedical and surgical aspects of captive reptile husbandry. Krieger, Melbourne, FL. 712 pp.

Gaulke, M. 1989. Zur biologie des bindenwaranes, unter berucksichtigung der paleogeographischen verbeitung und der phylogenetischen entwicklung der Varanidae. Cour. Forsch.-Inst. Senckenb. 112: 1–229.

———. 1991a. On the diet of the water monitor, *Varanus salvator*, in the Philippines. Mertensiella 2:143–153.

———. 1991b. Systematic relationships of the Philippine water monitors as compared with *Varanus s. salvator*, with a discussion of dispersal routes. Mertensiella 2:154–167.

———. 1994. Observations on the habitat selection of water monitors (*Varanus salvator*), p. 95, *In* Abstr. 2nd World Congres Herpet., Adelaide.

Gauthier, J., R. Estes, K. de Queiroz. 1988. A phylogenetic analysis of Lepidosauromorpha. Pp. 119–280 *In* R. Estes and G. Pregill (eds.) Phylogenetic Relationships of Lizard Families. Stanford U. Press.

Gray, J.E. 1827. A synopsis of the genera of saurian reptiles in which some new genera are indicated, and the others reviewed by actual examination. Philos. Mag.(n.s.) 2:54–58.

Green, B. 1972. Water losses of the sand goanna (*Varanus gouldii*) in its natural environment. Ecology 53:456–456.

Green, B., and D. King. 1978. Home ranges and activity patterns of the sand goanna *Varanus gouldii*. Austr. J. Wildl. Res. 5:417–424.

———, G. Dryden, K. Dryden. 1991. Field energetics of a large carnivorous lizard, *Varanus rosenbergi*. Oecologia 88:547–551.

———, D. King, M. Braysher, A. Saim. 1991. Thermoregulation, water turnover and energetics of free-living Komodo dragons, *Varanus komodoensis*. Comp. Biochem. Physiol. 99A:97–101.

———, M. McKelvey, P. Rissmiller. 1994. The breeding biology of *Varanus rosenbergi*, p. 102, *In* Abstr. 2nd World Cong. Herpet., Adelaide.

Greer, A. E. 1989. The biology and evolution of Australian lizards. Surrey Beatty & Sons, Chipping Norton, NSW. 248 pp.

Greer, G. C. 1994. Nest sites of royal pythons in west Africa. Rept. and Amphib Mag. July-Aug. 1994:45–55.

Gunther, A. 1861. On the anatomy of *Monitor niloticus* from western Africa, and of *Regenia albogularis*. Proc. Zool. Soc. London 1861:109–113.

Gupta, B. K. 1994. Conservation and commercial utilization of endangered monitor lizards, p. 198, *In* Abstr. 2nd World Congress Herpet., Adelaide.

Heger, N. A., and T. G. Heger. 1994. Thermoregulation, activity pattern, and home range variation in the large monitor lizard, *Varanus giganteus*, p. 115, *In* Abstr. 2nd World Congress Herpet., Adelaide.

Hooijer, D.H. 1972. *Varanus* (Reptilia, Sauria) from the Pleistocene of Timor. Zool. Mededelingen 47:445–448.

Horn, H.G. 1977. Notes on the systematics, places of discovery and keeping of *Varanus karlschmidti*. Salamandra 13:78–88.

———. 1978. Nachzucht von *Varanus gilleni* (Reptilia:Sauria: Varanidae). Salamandra 14:29–32.

———. 1980. Previously undiscovered details about *Varanus varius* based on observations in the field and in the terrarium. Salamandra 16:1–18.

———. 1985. Comments on the behavior of monitors: ritual combats of *V. komodoensis* Ouwens 1912 and *V. semiremex* Peters 1869, and the threat display phases of the ritual combat of *V. timornsis timorensis* (Gray 1831) and *V. t. similis* Mertens 1958. Salamandra 21:169-179.

———, and G. Peters. 1982. Some notes on the biology of the rough-neck monitor, *Varanus (Dendrovaranus) rudicollis* Gray. Salamandra 18:29–40.

———, and B. Schulz. 1977. *Varanus dumerilii*, as few know it. Das Aquar. 91:37–38.

———, and U. Schurer. 1978. Some notes concerning *Varanus glebopalma* (Mitchell, 1955). Salamandra 14:105–116.

———, and G. J. Visser. 1988. Observations on *Varanus giganteus* with some morphometric statements. Salamandra 24:102–118.

——— and ———. 1989. Review of reproduction of monitor lizards, *Varanus* spp. Int. Zoo Yearbook 28:140–150.

——— and ———. 1991. Basic data on the biology of monitors. Mertensiella 2:176–187.

Horner, J. R. 1987. Ecologic and behavioral implications derived from a dinosaur nesting site. *In* S. J. Czerkas and E. C. Olson (eds.), Dinosaurs Past and Present, vol. II, Seattle, Univ. Wash. Press.

Igolkin, V. A. no date. The propagation of reptiles in captivity: the grey monitor *Varanus griseus*. Priroda 9:95–96.

James, C. D., J. B. Losos, D. R. King. 1992. Reproductive biology and diets of goannas (Reptilia: Varanidae) from Australia. J. Herpet. 26:128–136.

Keirans, J., D. King, R. D. Sharrad. 1994. Two new species of ticks from Australian varanid lizards, p. 138, *In* Abstr. 2nd World Cong. Herpet., Adelaide.

Khan, M. A. R. 1988. A report on the survey of the biological and trade status of *Varanus bengalensis, Varanus flavescens,* and *Varanus salvator* in Bangladesh. Unpubl. report.

King, D. 1980. The thermal biology of free-living sand goannas (*Varanus gouldii*) in southern Australia. Copeia 1980:755–767.

———. 1991. The effect of body size on the ecology of varanid lizards. Mertensiella 2:204–210.

———, and M. King, P. Baverstock. 1991. A new phylogeny of the Varanidae. Mertensiella 2:211–219.

———, and B. Green. 1993. Goanna. The biology of varanid lizards. New South Wales Univ. Press, Kensington NSW, 102 pp.

———, and L. Rhodes. 1982. Sex ratio and breeding season of *Varanus acanthurus*. Copeia 1982:784–787.
King, M. 1990. Chromosomal and immunogenetic data: a new perspective on the origin of Australia's reptiles. *In* Cytogenetics of Amphibians and Reptiles: Advances in Life Sciences. Birkhausen Verlag, Basel.
———, and P. Horner. 1987. A new species of monitor (Platynota: Reptilia) from norther Australia and a note on the status of *Varanus acanthurus insulanicus* Mertens. The Beagle, Rec. No. Terr. Mus. Arts Sci. 4:73–79.
———, and D. King. 1975. Chromosomal evolution in the lizard genus *Varanus* (Reptilia). Austr. J. Biol. Sci. 28:89–108.
———, G. A. Mengpen, D. King. 1982. A pericentric inversion polymorphism and a ZZ/ZW sex-chromosome system in *Varanus acanthurus* analyzed by G- and C-banding and Ag staining. Genetica 58:39–45.
Klag, K., and H. Kantz. 1988. Bemerkungen zur haltung und fortpflazung von *Varanus bengalensis* im terrarium. Herpetofauna Weinstadt 10:21–24.
Kratzer, H. 1973. Observations on the incubation time of a clutch of eggs of *Varanus salvator*. Salamandra 9:27–33.
Krebs, U. 1991. Ethology and learning: from observations to semi-natural experiment. Mertensiella 2:220–232.
Lange, J. 1989. Observations on the Komodo monitor, *Varanus komodoensis* in the Zoo-Aquarium Berlin. Int. Zoo Yearbook 28:151–153.
Laurent, R. F. 1964. A new subspecies of *Varanus exanthematicus* (Sauria, Varanidae). Breviora 199:1–5.
Linnaeus, C. 1758. Systema naturae, 10th ed., vol. 1.
Lonnberg, E. 1903. On the adaptations to a molluscivorous diet in *Varanus niloticus*. Ark. Zool. 1:65–83.
Loop, M. S. 1976. Auto-shaping—a simple technique for teaching a lizard to perform a visual discrimination task. Copeia 1976:574–576.
Losos, J. B. and H. W. Greene. 1988. Ecological and evolutionary implications of diet in monitor lizards. Biol. J. Linn. Soc. 35:379–407.
Louw, G. N., B. A. Young, J. Bligh. 1976. Effect of thyroxine and noradrenaline on thermoregularion, cardiac rate and oxygen consumption in the monitor lizard *Varanus albigularis albigularis*. J. Therm. Biol. 1:189–193.
Luxmoore, R., and B. Groombridge. 1990. Asian monitor lizards. A review of distribution, status, and exploitation and trade in four selected species. World Conservation Monitoring Centre, Cambridge UK, 195 pp.
Makayev, V. M. 1982. Present condition and problems of conservation of the desert monitor lizard (*Varanus griseus*). Maochnie Osnovi Patsionalnhogo Ispolzovaniya Zhivotnoho Mira, Moscow, 36–42.
McCoid, M. J., and R. A. Hensley. 1991. Mating and combat in *Varanus indicus*. Herp. Rev. 22:16–17.
———, ———, G. Witteman. 1994. Factors in the decline of *Varanus indicus* on Guam, Mariana Islands. Herp. Rev. 25:60–61.
Merrem, B. 1820. Tentamen systematis amphibiorum. Marburg, Germany.

Mertens, R. 1942a. Die familie der warane (Varanidae). Ersten teil: Allgemeines. Abh. Senckenb. 462:1–116.

———. 1942b. Die familie der warane (Varanidae). Zweiter teil: der schadel. Abh. Senckenb. 465:117–234.

———. 1942c. Die familie der warane (Varanidae). Dritter teil: Taxonomie. Abh. Senckenb. 466:235–391.

———. 1951. A new lizard of the genus *Varanus* from New Guinea. Fieldiana Zool. 43:467–471.

———. 1954. Uber die rassen des wustenwarans (*Varanus griseus*). Sencken. biol. 35:353–357.

———. 1963. Helodermatidae, Varanidae, Lanthonotidae. Liste der rezenten amphibien und reptilien. Das Tierreich 79:1–26.

———. 1966. Ein neuer awergwaran aus Australien. Senckenb. biol. 47:437–441.

———. 1971. Ueber eine waransammlung aus dem ostlichen Neuguinea. Senckenb. Biol. 52:1–5.

Moehn, L. D. 1984. Courtship and copulation in the Timor monitor, *Varanus timorensis*. Herp. Rev. 15:14–16.

Muller, L. 1905. Der westafrikanische steppenwaran (*Varanus exanthematicus* Bosc). Bl. Aquar. Terrar. Kunde 16:266–276.

Murphy, J. B., and L. A. Mitchell. 1974. Ritualized combat behavior of the pygmy mulga monitor, *Varanus gilleni* (Sauria:Varanidae). Herpetologica 30:90–97.

Nagy, K. A. 1982. Energy requirements of free-living iguanid lizards. Pp. 49–59. *In* G. M. Burghardt and A. S. Rand (eds.) Iguanas of the World. Noyes, Park Ridge, NJ.

National Research Council, 1986. Nutrient requirements of cats. Rev. ed., Nat. Acad. Press, Washington, DC, 78pp.

Nutphrand, W. (no date). The monitor lizards of Thailand. Mitphadung Publ., Bangkok, 33pp.

Ouwens, P. A. 1912. On a large *Varanus* species from the island of Komodo. Bull. Jard. Bot. Buit. 2:1–3.

Pengilley, R. 1981. Notes on the biology of *Varanus spenceri* and *V. gouldii*, Barkly Tablelands, Northern Territory. Austr. J. Herp. 1:23–26.

Peters, U. 1969. Zum ersten mal in gefangenschaft: Eiablage und schlupf von *Varanus spenceri*. Aqua. Terr. 16:306–307.

———. 1970. Taronga Zoo hatches Spencer's monitors. Anim. Kingd. 73:30.

———. 1971a. first hatching of *Varanus spenceri* in captivity. Bull. Zoo. Mgmt. 3:17–18.

———. 1971b. Remarks on Mitchell's water monitor. Aquaterra Z. 8:75–77.

———. 1973. A note on the ecology of *Varanus (Odatria) storri*. Das Aquar. 53:462–463.

———. 1986. The successful rearing of *Varanus spenceri*. Das Aquarium 205.

Peters, W. 1870. Uber die afrikanischen warneidechsen, monitors, und ihre geographische verbreitung. Mber. Akad. Wiss. Berlin 1870:106–110.

Phillips, J.A. 1995. Movement patterns and density of *Varanus albigularis*. J. Herpet. 29:407–416.

———, and A. C. Alberts. 1994. Ecology of *Varanus albigularis*, p. 198, *In* Abstr. 2nd World Congress Herpet., Adelaide.

———. and G.C. Packar. 1994. Influence of temperature and moisture on eggs and embryos of the white-throated savannah monitor, *Varanus albigularis*: implications for conservation. Biol. Cons. 69:131–136.

Pianka, E. R. 1968. Notes on the biology of *Varanus eremius*. West. Austr. Nat. 11:39–44.

———. 1969. Notes on the biology of *Varanus caudolineatus* and *Varanus gilleni*. West Austr. Nat. 11:76–82.

———. 1970. Notes on the biology of *Varanus gouldii flavirufus*. West. Austr. Nat. 11:141–144.

———. 1971. Notes on the biology of *Varanus tristis*. West Austr. Nat. 11:180–181.

———. 1982. Observations on the ecology of *Varanus* in the Great Victoria Desert. West Austr. Nat. 15:1–8.

———. 1994a. Comparative ecology of *Varanus* in the Great Victoria Desert, p. 199, *In* Abstr. 2nd World Congress Herpet., Adelaide.

———. 1994b. Comparative ecology of *Varanus* in the Great Victorian Desert. Austr. J. Ecol. 19:395–408.

Pregill, G. K., J. A. Gauthier, H. W. Greene. 1986. The evolution of helodermatid squamates, with description of a new taxon and an overview of Varanoidea. Trans. San Diego Nat. Hist. Soc. 21:167–202.

Radford, L., and F. L. Paine. 1989. The reproduction and management of the Dumeril's monitor, *Varanus dumerilii*, at the Buffalo Zoo. Int. Zoo Yearbook 28:153–155.

Rao, R.J. and B.B. Sharma. 1994. Status and conservation of monitor lizards in India, p. 210, *In* Abstr. 2nd World Congress Herpet., Adelaide.

Rathgen, W. 1894. Der wustenwaran im terrarium. Natur and Haus 2:169–171.

Ratnam, J. 1994. Resource partitioning in a captive population of the water monitor *Varanus salvator*, p. 211, *In* Abstr. 2nd World Congress Herpet., Adelaide.

Rjumin, A.V. 1968. The ecology of the desert monitor (*Varanus griseus*) in southern Turkmenistan. *In* Herpetology of Middle Asia, Acad. Sci., Uzbek SSR, Tashkent, pp. 28–31.

Ruegg, R. 1974. The breeding of *Varanus timorensis similis*. Das Aquarium 62:360–363.

Sauterau, L., and P. de Bitter. 1980. Notes on the rearing and reproduction of the Timor monitor. Bull. Soc. Herp. Franc. 15:4–9.

Scherer, J. 1907. Eine jagd auf dem nilwaran am Senegal. Bl. Aquar. Terrar. Kunde 18:1–4.

Schmida, G. E. 1974. The short tailed monitor (*Varanus brevicauda*). Aqua. Terra. 27:390–394.

Schmutz, E., and H. G. Horn. 1986. The habitat of *Varanus timorensis timorensis*. Salamandra 22:147–156.

Schurer, U., and H. G. Horn. 1976. Observations of wild and captive Australian water monitors: *Varanus mertensi*. Salamandra 12:176–188.

Schwenk, K. 1985. On the occurrence, distribution, and functional significance of taste buds in lizards. Copeia 1985:99–101.

———. 1994. Why snakes have forked tongues. Science 263:1573–1577.

Shammakov, S. 1981. The reptiles of Turkmenistan. Acad. Nauk. Turkmen., 312 pp.

Shea, G.M. 1994. Three species of goanna occur in the Sydney Basin. Herpetofauna 24(2):14–18.

Shine, R. 1986. Food habits, habitats and reproductive biology of four sympatric species of varanid lizards in tropical Australia. Herpetologica 42:346–360.

Smith, H. C. 1931. The monitor lizards of Burma. J. Bombay Nat. Hist. Soc. 34:367–373.

Smith, M. A. 1932. Some notes on the monitors. J. Bombay Nat. Hist. Soc. 35:615–619.

———. 1935. The fauna of British India, vol. 2, Sauria:397–408. Taylor and Francis Ltd., London.

Sprackland, R. G. 1989. Captive maintenance of green tree monitors (*Varanus prasinus*) and their kin. No. Calif. Herp. Soc. Spec. Publ. 5:49–56.

———. 1991a. The origin and zoogeography of monitor lizards of the subgenus *Odatria* Gray (Sauria: Veranidae): a reevaluation. Mertensiella 2:240–252.

———. 1991b. Taxonomic review of the *Varanus prasinus* group with description of two new species. Mem. Queensland Mus. 30:561–576.

———. 1993. Rediscovery of a Solomon Islands monitor lizard (*Varanus indicus spinulosus*) Mertens, 1941. Vivarium 4(5):25–27.

———. 1994. Rediscovery and taxonomic review of *Varanus indicus spinulosus* Mertens, 1941. Herpetofauna 24(2):33–39.

Stanner, M. 1983. The etho-ecology of the desert monitor (*Varanus griseus*) in the sand dunes south of Holon, Israel. unpubl. thesis. Tel Aviv Univ.

———. 1991. Activity patterns of the desert monitor (*Varanus griseus*) in the southern coastal plain of Israel. Mertensiella 2:253–262.

———, and H. Mendelssohn. 1987. Sex ratio, population density and home range of the desert monitor (*Varanus griseus*) in the southern coastal plain of Israel. Amphib.-Rept. 8:153–164.

Stirnberg, E., and H. G. Horn. 1981. An unexpected new generation in a terrarium: *Varanus (Odatria) storri*. Salamandra 17:55–62.

Storr, G. H. 1966. Rediscovery and taxonomic status of the Australian lizard *Varanus primordius*. Copeia 1966:583–584.

———. 1980. The monitor lizards (genus *Varanus* Merrem, 1820) of Western Australia. Rec. West. Austr. Mus. 8:237–293.

———, L. A. Smith, R. E. Johnstone. 1983. Lizards of Western Australia II:Dragons and Monitors. West Austr. Mus., Perth, 113 pp.

Subba Rao, M. V., and K. Kameswara Rao. 1982. Feeding ecology of the Indian common monitor, *Varanus monitor*. 6th Ann. Rept. Symp. captive propagation and husbandry, Washington, DC.

Surahya, S. 1989. An anatomical study of the Komodo dragon and its position in animal systematics. Gadjah Mada Univ. Press, Yogyakarta, 2 vols.

Tasoulis, T. 1992. Nesting observations on the lace monitor *Varanus varius*. Herpetofauna 22:46–47.

Taylor, E. H. 1963. The lizards of Thailand. U. Kansas Sci. Bull. 44:687–1077.

Thompson, G. G., and P. I. Withers. 1992. Effects of body mass and temperature on standard metabolic rates for two Australian varanid lizards (*Varanus gouldii* and *V. panoptes*). Copeia 1992:343–350.

——— and ———. 1994. Standard metabolic rates of two small Australian varanid lizards (*Varanus caudolineatus* and *Varanus acanthurus*). Herpetol. 50:494–502.

Traeholt, C. 1994. The behavioral ecology of the Malaysian water monitor, *Varanus salvator*, suppl., *In* Abstr. 2nd World Congress Herpet., Adelaide, suppl.

Tsellarius, A. Y., and Y. G. Menshikov. 1994. Indirect communication and its role in the formation of social structure in *Varanus griseus* (Sauria). Russian J. Herpet. 1:121–132.

Twigg, L.E. 1988. A note on agonistic behavior in lace monitors *Varanus varius*. Herpetofauna 18:23–25.

Van Duinen, J. J. 1983. Breeding *Varanus exanthematicus albigularis*. Lacerta 42:12–14.

Vernet, R. 1982. Study of the ecology of *Varanus griseus* in the northwestern Sahara. Bull. Soc. Herp. Franc. 22:33–34.

Villamor, C. I. 1994. Morphometric gonadal and feeding characteristics and some insights on the conservation status of the water monitor (*Varanus salvator*), p. 272, *In* Abstr. 2nd World Congress Herpet., Adelaide.

Visser, G. J. 1981. Breeding the white-throated monitor *Varanus exanthematicus albigularis* at Rotterdam Zoo. Int. Zoo Yb. 21:87–91.

———. 1985. Notes on the reproductive biology of *Varanus (Empagusia) flavescens* (Hardwicke and Gray 1827) in the Rotterdam Zoo. Salamandra 21:161–168.

von During, M., and M. R. Miller. 1979. Sensory nerve endings of the skin and deeper structures, vol. 9, pp. 407–441, *In* Biology of the Reptilia, C. Gans, R. G. Northcutt, P. Ulinski eds., Academic Press.

Walsh, T., and R. Rosscoe. 1994. The history, husbandry, and breeding of komodo monitors at the National Zoological Park, p. 276, *In* Abstr. 2nd World Congres Herpet., Adelaide.

Webb, G. J. W. 1994. The links between wildlife conservation and sustainable use p. 267, *In* Abstr. 2nd world Congress Herpet., Adelaide.

Wever, E. G. 1978. The reptile ear. Princeton Univ. Press, NJ. 1023pp

Western, D. 1974. The distribution, density and biomass density of lizards in a semi-arid environment of northern Kenya. E. Afr. Wildl. Jour. 12:49–62.

Whitaker, R., and Z. Whitaker. 1978. Distribution and status of *Varanus salvator* in India and Sri Lanka. Herpet. Rev. 11:81–82.
Wicker, R. 1994. The captive breeding of *Varanus acanthurus* in the Zoological Garden Frankfurt, p. 284, *In* Abstr. 2nd World Congress Herpet., Adelaide.
Wikramanayake, E. D., and B. Green. 1987. Thermoregulatory influences on the ecology of two sympatric varanids in Sri Lanka. Biotropica 21:74–79.
———, and G.L. Dryden. 1988. The reproductive ecology of *Varanus indicus* on Guam. Herpetologica 44:338–344.
———, D. Marcellini, W. Ridwan. 1994. The thermal ecology of adult and juvenile komodo dragons, *Varanus komodoensis*, p. 284, *In* Abstr. 2nd World Congress Herpet., Adelaide.
Withers, P C., and G. G. Thompson. 1994. The relationship between body mass and metabolism for Australian goannas, p. 287, *In* Abstr. 2nd World Congress Herpet., Adelaide.
Yadgarov, T. Y. 1968. Materials on the ecology of the desert monitor (*Varanus griseus*) in the Surhandaria River Basin. *In* Herpet. of Middle Asia, Acad. Sci. Uzbek SSR, Tashkent, pp. 24–28.
Yang Datong, and Li Simin. 1987. A new species of *Varanus* from Yunnan, with morphological comparison between it and six other species from southeast Asia. Chinese Herp. Res. 1:60–63.
———, and Liu Wanzhao. 1994. Relationships among species groups of *Varanus* from southeastern Asia with description of a new species from Vietnam. Chinese Zool. Res. 15:11–15.
Yeboah, S. 1993. Aspects of the biology of two sympatric species of monitor lizards, *Varanus niloticus* and *Varanus exanthematicus* (Reptilia, Sauria) in Ghana. Afr. J. Ecol. 32:331–333.
Zimmerman, E. 1986. Breeding terrarium animals. TFH, Neptune, NJ, 384 pp.

SCIENTIFIC NAME INDEX

Acacia, 82, 125, 127
Amblyomma, 103
Amblyrhynchus cristatus, 46
Aponomma, 103
Artemisia monosperma, 75

Bufo marinus, 113

Casuarina, 125
Cherminotus, 14
Conolophis pallidus, 41
Crocodylus niloticus, 56
Cyclura nubila, 41

Dipsosaurus dorsalis, 42
Dracunculus, 104

Eimeria, 109
Entamoeba invadens, 104
Eucalyptus, 125

Gekko gecko, 31

Heloderma, 35
Homo sapiens, 86
Hydrosaurus giganteus, 125
Hydrosaurus gouldii, 124
Hydrosaurus marmoratus, 136
Hydrosaurus nuchalis, 136
Hydrosaurus salvator, 134

Iguana iguana, 31

Klassia, 104

Lacerta exanthematicus, 123
Lacerta monitor, 8, 115, 118, 131
Lacerta varia, 115, 141

Megalania prisca, 16
Monitor, 115
Monitor beccarii, 118
Monitor dumerilii, 122
Monitor flavescens, 123

Monitor indicus, 128
Monitor kalabeck, 121
Monitor nebulosus, 119
Monitor nigricans, 134
Monitor niloticus, 131
Monitor prasinus, 132
Monitor salvadorii, 134
Monitor salvator, 134
Monitor togianus, 136
Monitor tristis, 140
Monocercomonas varani, 104

Odatria ocellata, 116
Odatria punctata, 140

Physaloptera, 104
Plasmodium,, 104
Protocephalus, 104
Psammosaurus caspius, 128
Python regius, 114

Reptiliotrema, 104

Saniwa, 15
Saniwides mongoliensis, 14
Sauromalus, 46
Sceloporus, 104
Stellio salvator, 134

Tanqua, 105
Telmasaurus, 14
Triodia, 121
Tupinambis albigularis, 116
Tupinambis bengalensis, 118
Tupinambis bivittatus, 135
Tupinambis griseus, 127
Tupinambis indicus, 128
Tupinambis monitor, 14, 115

Varanus acanthurus, 3, 13, 14, 16, 18, 46, 49, 54, 66, 67, 70–72, 76, 82, 98, 99, 110
Varanus a. acanthurus, 24, 116, 144

Varanus a. brachyurus, 24, 116, 144
Varanus a. insulanicus, 24, 116, 144
Varanus albigularis, 8, 13, 15–17, 29, 32, 38, 39, 42, 48–50, 58, 60, 61, 63, 64, 66, 70–72, 74, 76, 94, 99, 110, 112, 141
Varanus a. albigularis, 22, 73, 75, 78, 83, 103, 116, 146
Varanus a. angolensis, 22, 117, 146
Varanus a. ionidesi, 22, 117, 146
Varanus a. microstictus, 22, 117, 146
Varanus baritji, 15, 18, 23, 66, 118, 144
Varanus beccarii, 17, 23, 29, 39, 44, 110, 118, 147
Varanus bengalensis, 8, 13–17, 25, 26, 29, 31, 36, 38, 39, 47–49, 53, 54, 56, 57, 59, 61–66, 68, 70, 71, 72, 76, 78–83, 86, 98, 99, 107–112, 175
Varanus b. bengalensis, 20, 73, 75, 118, 148
Varanus b. irrawadicus, 20, 119, 148
Varanus b. nebulosus, 20, 119, 148
Varanus b. vietnamensis, 20, 120, 148
Varanus bivittatus, 135
Varanus bogerti, 17, 23, 120, 147
Varanus brevicauda, 16, 18, 24, 49, 54, 66, 70, 99, 120, 145
Varanus caspius, 128
Varanus caudolineatus, 18, 24, 48, 54, 63, 66, 81, 121, 149
Varanus cumingi, 135
Varanus d. doreanus, 17, 21, 29, 110, 121, 171
Varanus d. finschi, 21, 122, 171
Varanus d. dumerilii, 3, 13, 16, 17, 20, 29, 61–63, 66, 69, 70, 72, 99, 110, 111, 122, 150

Varanus d. heteropholis, 20, 122, 150
Varanus eremius, 13, 16, 18, 23, 48, 54, 56, 66, 79, 81, 110, 123, 151
Varanus exanthematicus, 3, 8, 13, 14, 17, 22, 39, 48, 54, 65, 66, 69, 72, 75, 79, 81, 83, 89, 94, 109, 110, 112, 117, 123, 146
Varanus flavescens, 13–17, 20, 29, 36, 46, 55, 66, 68, 70, 72, 73, 76, 82, 86, 99, 107–112, 114, 123, 152
Varanus f. flavirufus, 13, 15, 16, 18, 22, 41, 44, 46, 48–50, 53, 54, 56, 62, 64, 66, 67, 70–72, 78, 80, 81–84, 99, 110, 124, 153
Varanus f. gouldii, 22, 124, 153
Varanus giganteus, 15, 16, 18, 21, 26, 46, 48, 54, 56, 62, 64, 66, 70, 72, 73, 79–81, 84, 86, 94, 99, 125, 154
Varanus gilleni, 13, 15, 16, 18, 24, 38, 39, 48, 50, 54, 61, 63, 64, 66, 70–72, 81, 99, 110, 125, 155
Varanus glauerti, 16, 18, 23, 54, 64, 69, 125, 155
Varanus glebopalma, 18, 23, 38, 56, 66, 79, 82, 126, 154
Varanus g. gouldii, 13–16, 18, 22, 48, 54, 56, 64, 66, 78, 82, 84, 86, 110, 126, 156
Varanus g. horni, 22, 127, 156
Varanus g. rubidus, 22, 127, 156
Varanus grayi, 131
Varanus griseus, 8, 13–17, 26, 31, 36, 39, 46, 56, 66, 69, 72, 79, 81, 82, 86, 94, 99, 107, 109–111
Varanus g. griseus, 4, 22, 54, 73, 78, 111, 127, 157
Varanus g. caspius, 22, 71, 78, 108, 128, 157
Varanus g. koniecznyi, 22, 55, 128, 157

Varanus heteropholis, 122
Varanus hofmanni, 15
Varanus indicus, 7, 13–17, 21, 39, 44, 50, 54, 61, 63, 72, 76, 109, 110, 113, 121, 128, 158, 173
Varanus indicus kalabeck, 121
Varanus irrawadicus, 119
Varanus jobiensis, 7, 15–17, 21, 29, 44, 50, 66, 110, 129, 172
Varanus kalabeck, 121
Varanus karlschmidti, 7, 129
Varanus kingorum, 15, 16, 18, 24, 129, 145
Varanus komodoensis, 4, 13, 15–17, 21, 25, 35, 38, 39, 44, 48, 50, 53–58, 61–67, 69–76, 79, 81, 82, 85, 86, 99, 107, 108, 110, 129, 160
Varanus kordensis, 132
Varanus manilensis, 136
Varanus mertensi, 14–16, 18, 21, 39, 55, 62, 64, 66, 70, 72, 78, 81, 82, 99, 110, 130, 151
Varanus microstictus, 117
Varanus mitchelli, 14–16, 18, 22, 55, 66, 72, 82, 110, 130, 149
Varanus nebulosus, 13, 119
Varanus niloticus, 4, 8, 13–17, 20, 31, 35, 39, 46, 55, 56, 62–64, 66, 69, 72, 75, 76, 79, 81, 83, 86, 89, 108–111, 131, 161
Varanus nuchalis, 136
Varanus olivaceus, 17, 20, 35, 36, 39, 47, 49, 53, 55, 58, 59, 61–67, 70, 71, 73–76, 78, 79, 81, 82, 85, 86, 99, 108, 110, 112, 131, 162, 175
Varanus ornatus, 128, 131
Varanus panoptes, 126–127
Varanus "pelewensis," 173
Varanus pilbarensis, 18, 24, 79, 132, 163
Varanus prasinus, 13, 14, 16, 17, 23, 38, 39, 44, 55, 64, 66, 67, 70, 72, 84, 99, 110, 118, 120, 132, 147, 173

Varanus primordius, 16, 18, 23, 50, 133, 145
Varanus punctatus, 140
Varanus rosenbergi, 13, 15, 16, 22, 41, 44–49, 55–59, 61, 63–66, 68, 70, 71, 73, 74, 76, 78–81, 86, 94, 110, 133, 153
Varanus rudicollis, 13, 14–17, 22, 29, 36, 39, 44, 55, 66, 70, 71, 99, 110, 111, 133, 164
Varanus salvadorii, 13, 14, 16, 20, 29, 39, 55, 110, 134, 159
Varanus salvator, 4, 8, 11, 13, 14–17, 29, 36, 39, 43, 46–48, 56, 62–66, 70–72, 75, 76, 79, 81–83, 99, 108–112, 173
Varanus s. salvator, 21, 73, 78, 134, 165
Varanus s. andamanensis, 21, 135, 165
Varanus s. bivittatus, 21, 135, 165
Varanus s. cumingi, 21, 108, 135, 165
Varanus s. marmoratus, 21, 86, 136, 165
Varanus s. nuchalis, 21, 136, 165
Varanus s. togianus, 21, 136, 165
Varanus scalaris, 18, 23, 82, 137, 167
Varanus scincus, 127
Varanus scutigerulus, 134
Varanus semiremex, 15, 18, 22, 46, 64, 66, 86, 137, 145
Varanus spenceri, 15, 16, 18, 21, 62, 64, 66, 70, 79, 86, 99, 137, 163
Varanus spinulosus, 17, 21, 138, 170
Varanus s. storri, 15, 18, 23, 55, 56, 66, 67, 70, 71, 99, 110, 138, 166
Varanus s. ocreatus, 23, 138, 166
Varanus telenestes, 17, 23, 139, 147
Varanus teriae, 17, 23, 139, 147

Varanus t. timorensis, 13, 15, 16, 18, 23, 58, 60, 61, 63, 66, 71, 72, 99, 110, 139, 167
Varanus t. similis, 23, 140, 167
Varanus togianus, 136
Varanus t. tristis, 13, 15, 16, 18, 23, 55, 56, 65, 66, 70, 72, 76, 79, 81, 82, 99, 110, 140, 168

Varanus t. orientalis, 23, 71, 140, 168
Varanus varius, 8, 13, 14–17, 21, 39, 47, 48, 55, 56, 59, 61–64, 66, 67, 70–72, 76–79, 81, 99, 110, 141, 166
Varanus vietnamensis, 120
Varanus vittatus, 134

Varanus yemenensis, 17, 22, 141, 169

GENERAL INDEX

acclimation to captivity, 89–90
accommodation of lens, 27
activity patterns, 76, 78–80
adductor mandibulae muscles, 30
aerobic capacity, 41–42
Africa, 1, 12, 14, 15, 74, 83, 111–112, 146, 161
aggression, 83–85
aigialosaurids, 14
air sacs, 41
albumin, 16, 68
allantois, 68
amino acid requirements, 95–96
amnion, 68
amniota, 9
ampullary crest, 31
anaerobic respiration, 41
Anguimorpha, 9
aorta, 43
arboreality, 39
Archosauria, 9
arginine vasotocin, 45
argus monitor (see Horn's monitor)
Asia, 1, 12, 15, 35, 110, 112–113, 148, 165
atria, 43
auditory tube, 30
Auffunberg, Walter, 1, 3, 4
Australia, 1, 3, 12, 15, 17, 78, 82, 110, 113, 144, 145, 149, 151, 153–156, 163, 166–168
Aves, 9

banded tree goanna (see spotted tree monitor, Australian)
Bangladesh, 75, 107, 112
basihyoid, 34
basilar membrane, 31
basking, 80
Beccari's monitor (see black tree monitor)
beetles (as food), 54–55

Bengal monitor, 118, 148
Benin, 112
bile, 37
biomass, 75
bipedalism, 38, 62, 83–84
birds (as food), 54–55
biting, 3, 85
black-headed monitor, 140, 168
black jungle monitor (see rough-necked monitor)
black-palmed rock monitor (see long-tailed rock monitor)
black-tailed monitor (see black-headed monitor)
black tree monitor, 118, 147
blood, 37
blue-nosed goanna (see Cape York tree monitor)
blue-tailed monitor, 121, 171
blunt-spined monitor, 133, 145
Bogert's tree monitor, 120, 147
Böhme, Wolfgang, 5, 7, 49
Borneo, 3
Bosc's monitor (see savannah monitor)
Boulenger, George, 8
breathing rate, 42
bronchi, 40–41
bulliwallah (see Mertens' water monitor)
bungarra, 124, 153
Burden, W. Douglas, 4
Burma, 4, 37
burrows, 44, 58, 80, 83, 91

caging, 90–94
calcium deficiency, 102
Cameroon, 111
cannibalism, 56, 86
Cape York tree monitor, 139, 147
captive breeding, 2, 98–100, 175
capture, 89, 109
carrion, in diet, 53, 56

Caspian monitor (see desert monitors, eastern)
cavum arteriosum, 42–43
cavum pulmonale, 42–43
cavum venosum, 42–43
centipedes (as food), 54–55
ceratobranchial, 34
ceratohyoid, 34
Chad, 111
character states, 9
Chelonia, 9
choanae, 25
cholesterol, 37
chorion, 68
choroid, 28
chromosomes, 15, 76–77
ciliary muscles, 27–28
circulatory system, 42–43
CITES, 107
cladistics, 11
cladograms, 12–18
climbing, 38
cloaca, 37, 45–46, 52, 58
cloacal pouch, 52
clouded monitor, 119, 148
clutch size, 65–66
cochlear duct, 30–31
collecting duct, 45
colon, 37, 51
color, skin, 69, 82
columnella, 30
compliance, lung, 41
cones (photoreceptors), 27
conjunctiva, 26
conservation, 107–114
conus papillaris, 27–28
coprodaeum, 45–46, 50, 51
copulation, 51, 57–61
core activity areas, 74
cornea, 26, 28
costal breathing, 41
courtship, 57–61
Cowles, Raymond, 4

crabs (as food), 54–55
Cretaceous, 1, 14
crocodile monitor (see Papua monitor)
crocodiles, 9
crypsis, 83

daily activity, 78–80
Daudin, Francois, 8
defense, 83–85
desert monitors
    eastern, 128, 157
    rusty, 123, 151
    Thar, 128, 157
    western, 127, 157
detoxification, by liver, 37
diet, artificial, 96
    natural, 53–56
digestive efficiency, 38
digestive time, 38
Dinosauria, 9
diseases, 100–105
Ditmars, Raymond, 4
dolichosaurids, 14
dominance hierarchy, 74, 83
dreamtime, 3
drinking, 35
ductus deferens, 49–50
Dumeril's monitor, 122, 150
duodenum, 37

ear, 27, 29–32
education, 114
eggs (monitor), 51–52, 65–68
eggs (as food), 54–56
egg size/mass, 65–66
egg-tooth, 69
electrolyte homeostasis, 44
embryo, 68–69
emerald monitor (see green tree monitor)
endangered species, 107–108
endolymph, 31
endolymphatic sac, 30
energetics, 47–49
Eocene, 15
epidermal glands, 61

epidermal senses, 32
epididymis, 49–50
equilibrium sense, 30
esophagus, 35, 37
Europe, 15
evaporative water loss, 44
external ear, 29
extrastapes, 30
eye, 26–28
eyelids, 26, 28

fat bodies, 37, 49
faveoli, 41
fecal scent, 74
feces, 46, 74
feeding, captive, 94–98
femoral pores, 9
field metabolic rates, 48
Finsch's monitor, 122, 171
fish (as food), 54–55
flight distance, 83
flute monitor (see rough-necked monitor)
follicles, ovarian, 51–52
food consumption, 57
foraging, 56–57, 78
force feeding, 97–98
fossils, 14–18
fovea, 27–28
France, 108
freckled monitor, 140, 168
frogs (as food), 54–55
fruit (as food), 5, 55

gastric pellets, 38
gastrointestinal disease, 101–102
Gaulke, Maren, 5
glands
    epidermal, 32, 61
    external nasal (salt), 25, 46–47
    femoral, 9
    Harderian, 29
    preanal, 9
    proctodeal, 61
    salivary, 35
        infralabial, 35
        mandibular, 35

    shell, 52
Glauert's monitor, 125, 155
glomerular filtrate, 45
glomeruli, 45
glottis, 33
glucose, 47
glycogen, 37
goannas, 3
Gould's monitor, 126, 156
gout, 102
Gray, John E., 8
Gray's monitor, 131, 162
Green, Brian, 5
green tree monitor, 132, 147
growth rate, 70–71
gular fluttering, 81
gular inflation, 85
Gunther, Albert, 4

habitats, 82–83
habitat destruction, 111, 113
hair cells, 31
handling, 94
harlequin monitor (see rough-necked monitor)
hatching, egg, 68–69
hatchlings, 69–70
health, 89
hearing, 31
heart, 37, 42–43
heart rate, 42
heath monitor (see Rosenberg's monitor)
heating, cage, 92–93
hemibaculum, 49
hemipenes, 13, 49–51, 58
Hindu lore, 1
home range, 73–74
Horn's monitor, 127, 156
hyoid apparatus, 34–35

Iguania, 9
ileum, 37
incubation period, 66–68
India, 37, 75, 79, 82, 111–114, 152
Indonesia, 1, 108, 112, 150, 160

infralabial gland, 35
inner ear, 30–31
insects (as food), 54–55
intelligence, 86–87
intercostal muscles, 41
intestines, 36
iris, 27
isopods (as food), 54–55
Israel, 75, 78
Italy, 108
IUCN Red List, 108

Jacobson's organ, 25
Japan, 108
jaws, 35–36

kalabeck monitor (*see* blue-tailed monitor)
Karl Schmidt's monitor (*see* peach-throated monitor)
Kenya, 75
keys, taxonomic, 20–24
kidney, 37, 40, 44–45, 51
King, Dennis, 5
Kings' monitor, 129, 145
Komain monitor (*see* water monitors, Asian)
Komodo Island, 4
Komodo monitor, 129, 160
Koniecznyi's monitor (*see* desert monitors, eastern)

lace monitor, 141, 166
lactic acid, 41, 43
lamellated skin receptors, 32
lateral undulation, 38
Laurasia, 14
legislation, 113
lens, 27–28
lens pad, 28
lepidopterans (as food), 54–55
Lepidosauria, 9
Lesser Sunda Islands, 1, 4
lights, cage, 91–92
line-tailed pygmy monitor (*see* stripe-tailed monitor)
lingual process, 34

Linnaeus, 8
liver, 37
lizards (as food), 54–55
locomotion, 38–39
longevity, 72
long-tailed rock monitor, 126, 154
Loonberg, Eimar, 4
lungs, 10, 40–41
Luzon, 5

maculae, 31
Malaysia, 3, 112, 150, 164
male combat, 62–63
Mali, 111
mammals, 9, 41, 46, 49
  (as food), 54–55
mandibular gland, 35
mangrove monitor, 128, 158
Marianas, 113
mating (*see* copulation)
mating systems, 63
Merrem, Blasius, 8
Mertens, Robert, 3, 4, 7, 8
Mertens' water monitor, 130, 151
metabolic rate, 48
Micronesia, 110
middle ear, 30
millipedes (as food), 54–55
Miocene, 15, 16, 19
Mitchell's water monitor, 130, 149
mollusks (as food), 54–55
Mongolia, 14
monitor, origin of name, 8
monogamy, 63
Montana, 15
mournful goanna (*see* freckled monitor)
mulga goanna (*see* pygmy mulga monitor)
Muller, Lorenz, 4
myoglobin, 43

Namibia, 38, 71, 78, 83
nares
  external, 25
  internal, 25
nasal capsule, 25, 47

nasal gland, 25, 46–47
nematodes, 104
nephrons, 46
nests, 64–65
New Guinea, 108, 112, 147, 159, 171, 172
nictitating membrane, 26
Nigeria, 111–112
Nile monitor, 131, 161
nitrogenous wastes, 44
nocturnal activity, 79
North America, 1, 14, 15
nostrils, 20
nutrition, 95–96

obesity, 97, 102
ocellate ridge-tailed monitor (*see* ridge-tailed monitor)
olfactory bulb, 25
olfactory chamber, 25
olfactory membrane, 25
olfactory nerve, 28
optic nerve, 28
ora (*see* Komodo monitor)
organ of Corti, 31
orthopterans (as food), 54–55
osteoderms, 9, 107
otic capsule, 30–31
Ouwens, P. A., 4
ovaries, 51–52
oviduct, 51–52
oviposition, 64–65
ovulation, 52
oxygen use, 40–42

Pakistan, 75, 76, 79, 82
pancreas, 37
Pangea, 14
papilla basilaris, 31
papillae, sensory, 32
Papua monitor, 134, 159
parasites, 102–105
peach-throated monitor, 129, 172
perentie, 125, 154
pericardium, 42
perilymph, 31
perilymphatic space, 30

Peters, Wilhelm, 4
pharynx, 30
phenetics, 11
pheromones, 61
Philippines, 75, 83, 108, 112, 162
photoperiod, 91
phylogeny, 10–18
Pianka, Eric, 5
Pilbara monitor, 132, 163
Pleistocene, 16, 17
pleural cavity, 41
population density, 74–75
population dynamics, 73–76
poryphasmata, 50
poultice, 3
preanal glands, 9
predators, 86
prey types, 54–55
proctodaeum, 46
proctodeal glands, 61
proteins, 47
protist parasites, 104
pseudotemporalis muscle, 30
pulmonary arteries, 43
pulmonary veins, 43
pygmy mulga monitor, 125, 155
pygmy rock monitor (see Kings' monitor)

quadrate bone, 30

racehorse goanna (see Gould's monitor or freckled monitor)
radiations, 12, 17–19
range maps
 genus, 2
 species, 143–172
Rathgen, W., 4
rectum, 37, 46, 50, 51
renal artery, 45
reproductive anatomy, 49–52
reproductive behavior, 55–70
reproductive maturity, 71
reproductive strategy, 70
reptilia, 9
respiration, 40–42
respiratory disease, 101
retinal tunic, 27–28

ridge-tailed monitors, 116, 144
ritual combat, 62–63
rock leguaan (see white-throated monitor, South African)
Rosenberg's monitor, 133, 153
Rossel Island monitor, 139, 147
rough-necked monitor, 133, 164
round window, 30
running, 38
rusty desert monitor, 123, 151
rusty monitor, 137, 145

saccule, 30–31
safety (captives), 90–91, 97
salivary glands, 35
salt balance, 46–47
salt glands, 45–47
Salvadori's monitor (see Papua monitor)
sand monitor, 124, 153
Sarawak forest monitor, 122, 150
Sauria, 9
savannah monitor, 123, 146
scales, 9, 20, 32
Scherer, J., 4
scleral cartilage, 28
scleral ossicles, 27–28
scleral tunic, 27
scorpions (as food), 54–55
seasonal activity, 76, 78
Seba, Albert, 4
semicircular canals, ducts, 30–31
Senegal, 4
Sepik monitor (see peach-throated monitor)
sex ratio, 76
short-tailed monitor, 120, 145
sinus venosus, 43
size range, 1
skeleton, 10
skin diseases, 101
skin trade, 3, 107–109
skull, 9–11
sleep, 79–80
smell, 25–26
Smith, H.C., 4
snails, 53

snakes
 (as food), 35–39
 (as predators), 86
sodium, 44–47
Solomons, 1, 108, 112, 170
Solomons keeled monitor, 138, 170
sound sensitivity, 31–32
South America, 14
species
 number of, 1
 threatened, 107–108
speed, 38
Spencer's monitor, 137, 163
spiders (as food), 54–55
spotted tree monitor, Australian, 137, 167
spotted tree monitor, New Guinea, 140, 167
Sprackland, R.G., 5
Squamata, 9
Sri Lanka, 2, 76, 82
standard metabolic rate, 48
stapes, 30
stereocilia, 31
stereopsis, 26
sternum, 25
stomach, 36, 37, 56
stomatitis, 101–102
Storr's monitor, 138, 166
stripe-tailed monitor, 121, 149
subgenera, 8
sublingual plicae, 25
Sudan, 111–112
sulcus spermaticus, 49–50
survivorship, 75
swallowing, 34–35
swimming, 39
sympatry, 1
symptoms of illness, 100
systole of heart, 42

Tadjikistan, 75
tail slapping, 84
Tanzania, 112
tarsus, 26, 28
taste sense, 33
taxonomy, 7–10, 115, 173
teeth, 35–36

temperature, activity, 81
termitaria, 64, 69
testes, 49–50
Thailand, 3, 76, 112
thermoregulation, 80–82
ticks, 102–104
Timor, 17
Timor monitor, 139, 167
Togo, 112
tongue, 25, 33–35
tongue sheath, 33
touch, sense of, 32
trachea, 30, 40–41
trade, live, 109–110
trade, skins, 107–109
tree crocodile (*see* peach-throated monitor)
tree leguaan (*see* white-throated monitor)
tree monitors
    Australian spotted, 137, 167
    black, 118, 147
    Bogert's, 120, 147
    Cape York, 139, 147
    green, 132, 147
    New Guinea spotted, 140, 167
Turkmenistan, 78, 107

tympanic cavity, 30
tympanic membrane, 29–30

underwater activity, 39
ureter, 40, 45–46
uric acid, 44
urinary papilla, 46
urine, 44–45
urodaeum, 45–46, 49–51
U.S.A., 108
uterus, 51–52
utricle, 30–31
Uzbekistan, 75, 78

Varanidae, 9
Varanoidea, 9, 14
venom, 35
vent, 37, 46, 51, 57
vestibular apparatus, 31
vestibulum, 25
Vietnam, 112
Vietnam monitor, 120, 148
vision, 26–27
visual field, 26–27
vitamins, 95–97
vitellogenesis, 52
vitreous body, 27–28
vomeronasal organ, 25–26, 33

walking, 38
"waran," 8, 115
water, for captives, 98
water balance, 44–47
water monitors
    Andaman Islands, 135, 165
    Asian, 134, 165
    Cuming's, 135, 165
    marbled, 136, 165
    Mertens', 130, 151
    Mitchell's, 130, 149
    Negros, 136, 165
    Togian, 136, 165
White's monitor, 116, 144
white-throated monitors, 116–118, 146
worms, 104–105

yellow monitor, 123, 152
yellow-spotted monitor, 127, 156
Yemen monitor, 140, 169
yolk, 52, 68
yolk sac, 68
yolk stalk, 68
Yunnan monitor, 119, 148

Zambia, 112